Foods and Packaging Materials — Chemical Interactions

Foods and Packaging Materials — Chemical Interactions

Edited by

Paul Ackermann

Tetra Pak (Research) GmbH, Stuttgart, Germany

Margaretha Jägerstad

Chemical Center, Lund Institute of Technology/University of Lund, Lund, Sweden

Thomas Ohlsson

SIK — The Swedish Institute for Food Research, Göteborg, Sweden

THE ROYAL
SOCIETY OF
CHEMISTRY

The Proceedings of an International Symposium on Interaction: Foods–Food Packaging Materials, held on 8–10 June 1994, in Lund. The Symposium was arranged by The Lund Institute of Technology, Lund University, and SIK, The Swedish Institute of Food Research, Göteborg.

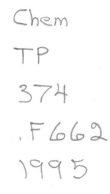
Special Publication No. 162

ISBN 0-85404-720-4

A catalogue record of this book is available from the British Library.

Published by The Royal Society of Chemistry,
Thomas Graham House, Science Park, Cambridge
CB4 4WF

Printed by Bookcraft (Bath) Ltd

Preface

Food and packaging interactions can be defined as chemical and/or physical reactions between a food, its package, and the environment, which alter the composition, quality, or physical properties of the food and/or the package. Interest in such interactions has increased during recent decades as a consequence of higher demands on food quality protection by packaging, the rapid development of new packaging materials and packaging technologies, new uses for packages, and concern for the environmental impact of packaging.

One of the problems associated with the use of polymer packages is the interaction between the package and its contents. Constituents from the packaging materials might be transferred into the food. This might cause off-flavour or be of toxicological concern. Such interaction, identified as migration, is generally dealt with by food contact material regulations. The opposite might also occur, i.e. food components may be sorbed by the packaging material. Depending on what type of substances that are sorbed, this phenomenon can give rise to various problems. The package itself and its properties might be altered, which can lead to the package not being able to maintain its protective role for the product. Sorption of food components by the package can also affect the foodstuff more directly. For instance, if key flavour compounds are lost to the packaging material, so-called flavour scalping, the aroma intensity decreases.

Scientific literature on food and packaging interactions is widely spread under various topics, often not easily accessible, and it seems to be difficult to obtain an overview of the current status of knowledge and research. Comprehensive scientific conferences have rarely been arranged. During the last decade, two meetings on this subject have been organized in the United States by Professor Joseph Hotchkiss of Cornell University. The Proceedings have been published by the American Chemical Society (ACS Symposium Series 365 and 473).

The symposium "Interactions: Foods - Food Packaging Material", held on June 8-10 at Lund University, Sweden, was the first scientific meeting directed towards this subject that has been organized in Europe. An important goal of this conference was to bring together active research groups from different parts of the world, to present their current knowledge and ongoing research, in order to provide an opportunity for the exchange of experience and to reveal needs for future research.

The presentations made and discussions held during this symposium focused on chemical interactions, physical interactions, process-induced interactions, interactions with new packaging materials, and active packaging. Migration in the context of regulations governing the contact of food with packaging materials has not been dealt with intentionally, since this topic is extensively covered regularly in a series of specific conferences.

The present monograph   covers many aspects, presenting both the state of the art and on-going research activities within the field, including the development of new techniques for studying the interactions between foods and their packaging materials. The objective of this publication is to give not only an overview of interaction phenomena in food packaging, but to help in the understanding of these phenomena and their consequences. Thus, it may also contribute to the improvement of food quality and to finding technical solutions in those cases where food and packaging interactions are responsible for adverse effects on food quality.

We wish to thank all the authors who contributed to the success of the symposium through their excellent presentations, and those who assisted in the making of this book.

Paul Ackermann
Tetra Pak Research GmbH, Stuttgart, Germany

Margaretha Jägerstad
Lund Institute of Technology, Lund University, Lund, Sweden

Thomas Ohlsson
The Swedish Institute for Food Research, Gothenburg, Sweden

Contents

List of Contributors

Ahvenainen, R.

VTT Biotechnology and Food Research
P.O Box 1500
SF-02044 VTT
Finland

Arnesen, A.K.

Matforsk - Norwegian Food Research Institute
Osloveien 1
N-1430 Ås
Norway

Axelsson-Larsson, L. *

Packforsk
P.O. Box 9
S-16493 Kista
Sweden

Bergqvist, A.-K.

STFI, Swedish Pulp and Paper Research Institute
P.O Box 5604
S-11486 Stockholm
Sweden

Bergslien, H.

Norconserv
Institute of Fish Processing and Preservation Technology
P.O. Box 327
N-4001 Stavanger
Norway

Bizet, C. *

Laboratoire de Physiochimie et Génie Alimentaires
Ecole Nationale Supérieure d'Agronomie et des Industries Alimentaires
Institut National Polytechnique de Lorraine
2 Avenue de la forêt de Haye - B.P 172
F-54505 Vandoeuvre-lès-Nancy Cedex
France

Bjerkeng, B. *

Norconserv
Institute of Fish Processing and Preservation Technology
P.O. Box 327
N-4001 Stavanger
Norway

Björklund-Jansson, M. *

STFI, Swedish Pulp and Paper Research Institute
P.O Box 5604
S-11486 Stockholm
Sweden

Day, B.P.F. *

Department of Product and Packaging Technology
Campden Food and Drink Research Association
Chipping Campden
Gloucestershire GL55 6LD
England

Debeaufort, F.

Ensbana
Laboratoire de Génie des Procédés Alimentaires et Biotechnologiques
Université de Bourgogne
1 Esplanade Erasme
F-21000 Dijon
France

Desobry, S. *

Laboratoire de Physiochimie et Génie Alimentaires
Ecole Nationale Supérieure d'Agronomie et des Industries Alimentaires
Institut National Polytechnique de Lorraine
2 Avenue de la forêt de Haye - B.P 172
F-54505 Vandoeuvre-lès-Nancy Cedex
France

Erlandsson, B.

Department of Nuclear Physics
University of Lund
Sölvegatan 14
S-22362 Lund
Sweden

Ewender, J.

Fraunhofer-Institut ILV
Schragenhofstrasse 35
D-80992 Munich
Germany

Feigenbaum, A. *

Institut National de la Recherche Agronomique
23 Rue Clément Ader
F-51100 Reims
France

Franz, R. *

Fraunhofer-Institut ILV
Schragenhofstrasse 35
D-80992 Munich
Germany

Giacin, J.R. *

School of Packaging
Michigan State University
East Lansing,
MI 48824-1223
USA

Gontard, N.

Ensia-Siarc
Food Engineering and Technology Unit
B.P. 5035
F-34090 Montpellier Cedex 1
France

Guilbert, S. *

Cirad-Sar
Food Engineering and Technology Unit
B.P. 5035
F-34090 Montpellier Cedex 1
France

Hardy, J.   Laboratoire de Physiochimie et Génie Alimentaires
                                 Ecole Nationale Supérieure d'Agronomie et des Industries Alimentaires
                                 Institut National Polytechnique de Lorraine
                                 2 Avenue de la forêt de Haye - B.P 172
                                 F-54505 Vandoeuvre-lès-Nancy Cedex
                                 France

Harmati, Z. *   Department Packaging/Paper
                                 Swiss Federal Laboratories for Materials Testing and Research
                                 CH-9001 St. Gallen
                                 Switzerland

Haugdal, J.   Matforsk - Norwegian Food Research Institute
                                 Osloveien 1
                                 N-1430 Ås
                                 Norway

Hellborg, R.   Department of Nuclear Physics
                                 University of Lund
                                 Sölvegatan 14
                                 S-22362 Lund
                                 Sweden

Hermansson, C.   SIK, the Swedish Institute for Food Research
                                 P.O Box 5401
                                 S-40229 Gothenburg
                                 Sweden

Hildingsson, I. *   Department of Chemical Engineering II
                                 Chemical Center
                                 Lund Institute of Technology
                                 P.O. Box 124
                                 S-22100 Lund
                                 Sweden

Hotchkiss, J.H. *   Institute of Food Science
                                 Food Science Department
                                 Cornell University
                                 Stocking Hall
                                 Ithaca
                                 NY 14853-7201
                                 USA

Hurme, E.   VTT Biotechnology and Food Research
                                 P.O Box 1500
                                 SF-02044 VTT
                                 Finland

Ishitani, T. *   Japan International Research Center for Agricultural Sciences
                                 Ministry of Agriculture, Forestry and Fisheries
                                 Tsukuba
                                 Japan

Janssens, J.L.G.M.   Wageningen Agricultural University
Department of Food Science
P.O. Box 8129
NL-6700 EV Wageningen
Netherlands

Johansson, F. *   SIK, the Swedish Institute for Food Research
P.O Box 5401
S-40229 Gothenburg
Sweden

Jägerstad, M.   Department of Applied Nutrition and Food Chemistry
Chemical Center
Lund Institute of Technology
P.O. Box 124
S-22100 Lund
Sweden

Latva-Kala, K.   VTT Biotechnology and Food Research
P.O Box 1500
SF-02044 VTT
Finland

Lea, P.   Matforsk - Norwegian Food Research Institute
Osloveien 1
N-1430 Ås
Norway

Leufvén, A. *   SIK, the Swedish Institute for Food Research
P.O Box 5401
S-40229 Gothenburg
Sweden

Lindberg, B. *   The Danish Packaging and Transportation Research Institute
Danish Technological Institute
Gregersensvej
P.O. Box 141
DK-2630 Taastrup
Denmark

Lindner-Steinert, A.   Fraunhofer-Institut ILV
Schragenhofstrasse 35
D-80992 Munich
Germany

Linssen, J.P.H. *   Wageningen Agricultural University
Department of Food Science
P.O. Box 8129
NL-6700 EV Wageningen
Netherlands

Mannheim, C.H.

Department of Food Engineering and Biotechnology
Technion - Israel Institute of Technology
Haifa 32000
Israel

Marque, D.

Institut de la Recherche Agronomique
LNSA
F-78532 Jouy en Josas
France

Miltz, J. *

Department of Food Engineering and Biotechnology
Technion - Israel Institute of Technology
Haifa 32000
Israel

Nicoli, P.

Distam
Department of Food Science and Microbiology
University of Milan
Via Celoria 2
I-20133 Milan
Italy

Nielsen, T. *

Department of Applied Nutrition and Food Chemistry
Chemical Center
Lund Institute of Technology
P.O. Box 124
S-22100 Lund
Sweden

Nilvebrant, N.-O.

STFI, Swedish Pulp and Paper Research Institute
P.O Box 5604
S-11486 Stockholm
Sweden

Novak, G.

Laboratoire de Physiochimie et Génie Alimentaires
Ecole Nationale Supérieure d'Agronomie et des Industries Alimentaires
Institut National Polytechnique de Lorraine
2 Avenue de la forêt de Haye - B.P 172
F-54505 Vandoeuvre-lès-Nancy Cedex
France

Olafsson, G. *

Department of Applied Nutrition and Food Chemistry
Chemical Center
Lund Institute of Technology
P.O. Box 124
S-22100 Lund
Sweden

Paik, J.S. *

Department of Food Science
University of Delaware
226 Alison Hall
Newark
DE 19716
USA

Passy, N.   Department of Food Engineering and Biotechnology
Technion - Israel Institute of Technology
Haifa 32000
Israel

Pieper, G. *   Tetra Pak (Research) GmbH
Waldburgstrasse 79
D-70563 Stuttgart
Germany

Piergiovanni, L. *   Distam
Department of Food Science and Microbiology
University of Milan
Via Celoria 2
I-20133 Milan
Italy

Piringer, O. *   Fraunhofer-Institut ILV
Schragenhofstrasse 35
D-80992 Munich
Germany

Randell, K. *   VTT Biotechnology and Food Research
P.O Box 1500
SF-02044 VTT
Finland

Reitsma, J.C.E.   Wageningen Agricultural University
Department of Food Science
P.O. Box 8129
NL-6700 EV Wageningen
Netherlands

Riquet, A.M.   Institut de la Recherche Agronomique
LNSA
F-78532 Jouy en Josas
France

Roozen, J.P.   Wageningen Agricultural University
Department of Food Science
P.O. Box 8129
NL-6700 EV Wageningen
Netherlands

Rosnes, J.T.   Norconserv
Institute of Fish Processing and Preservation Technology
P.O. Box 327
N-4001 Stavanger
Norway

Rüter, M.   Fraunhofer-Institut ILV
Schragenhofstrasse 35
D-80992 Munich
Germany

Sivertsvik, M.   Norconserv
                                  Institute of Fish Processing and Preservation Technology
                                  P.O. Box 327
                                  N-4001 Stavanger
                                  Norway

Skog, G.   Radiocarbon Dating Laboratory
                                  Department of Quaternary Geology
                                  University of Lund
                                  Tornavägen 13
                                  S-22362 Lund
                                  Sweden

Sørheim, O. *   Matforsk - Norwegian Food Research Institute
                                  Osloveien 1
                                  N-1430 Ås
                                  Norway

Stenström, K. *   Department of Nuclear Physics
                                  University of Lund
                                  Sölvegatan 14
                                  S-22362 Lund
                                  Sweden

Stöllman, U. *   SIK, the Swedish Institute for Food Research
                                  P.O Box 5401
                                  S-40229 Gothenburg
                                  Sweden

Tesson, N.   Ensbana
                                  Laboratoire de Génie des Procédés Alimentaires et Biotechnologiques
                                  Université de Bourgogne
                                  1 Esplanade Erasme
                                  F-21000 Dijon
                                  France

Tinelli, L.   Distam
                                  Department of Food Science and Microbiology
                                  University of Milan
                                  Via Celoria 2
                                  I-20133 Milan
                                  Italy

Törnell, B.   Department of Chemical Engineering II
                                  Chemical Center
                                  Lund Institute of Technology
                                  P.O. Box 124
                                  S-22100 Lund
                                  Sweden

Vestrucci, G.

CSI
Applied Research Centre
Montedison s.p.a.
Via Lombardia 20
I-20021 Bollate (Milan)
Italy

Voilley, A. *

Ensbana
Laboratoire de Génie des Procédés Alimentaires et Biotechnologiques
Université de Bourgogne
1 Esplanade Erasme
F-21000 Dijon
France

Wiebert, A.

Department of Nuclear Physics
University of Lund
Sölvegatan 14
S-22362 Lund
Sweden

* Principal authors

Chemical Interactions—General Approach

Overview on Chemical Interactions between Food and Packaging Materials

Joseph H. Hotchkiss

INSTITUTE OF FOOD SCIENCE, STOCKING HALL, CORNELL UNIVERSITY, ITHACA, NY 14853, USA

ABSTRACT

Food-package interactions can be defined as chemical and/or physical reactions between a food, its package, and the environment which alter the composition, quality, or physical properties of the food and/or package. Interest in such interactions has heightened as the change from more inert package materials to more reactive materials has occurred.

Most research has focused on the adverse effects of such interactions; changes in flavor due to aroma absorption by polymeric packaging, for example. Recently, interest in how such interactions might improve food quality has increased. The selective absorption of undesirable aromas and use of antimicrobial polymer materials are examples.

Several methods have been used to study interactions. Basic studies have modeled interactions with the objective of determining the fundamental nature of the interaction. Inverse gas chromatography is one such method (1). The absorption of aromas by polymeric materials using mass gain or direct extraction has been used (2). Permeation of films by aromas has also been investigated (3). Others have quantified aroma losses from foods or model foods (4). Some studies have relied on sensory analyses (5). Each of these studies has advantages and disadvantages.

We have recently combined the qualitative nature of gas chromatography-mass spectrometry with the sensitivity of the human nose (i.e. olfactometry) to understand these interactions. These data suggest that olfactometry can provide quantitative information on how these interactions affect food aroma. Data from the study of orange juice and oxidized polyethylene show the practical importance of understanding the sensory properties of these interactions.

1 INTRODUCTION

Even glass and metal containers are not completely inert with respect to foods. However, the large increase in the use of plastic and composite materials for packaging has increased the degree to which packaging directly affects food. In addition, recent interest in designing packaging that directly interacts with a food product and its environment to produce a desirable functional result has increased.

In general, food-package interactions can be divided into four types:

1. Transfer of package components to the product (migration)- This can result in safety concerns and flavor degradation. Transfer of desirable functional components such as antimicrobial agents may be beneficial.

2. Transfer of product components to the package (scalping)- Transfer of desirable aromas from food to packaging can result in flavor alteration and/or loss of package performance. Sorption of undesirable flavors or reduction in O_2 content of a package could be beneficial.

3. Transfer of product components through the package to the environment (egress permeation)- Loss of aroma-flavor volatiles, CO_2, or H_2O can result in changes in food quality.

4. Transfer of environmental components through the package to the product (ingress permeation)- Ingress of O_2, H_2O, light, or undesirable odors or toxicants can be detrimental. Packaging materials which interrupt this process (e.g., oxygen interceptors) can be beneficial.

2 UNDESIRABLE EFFECTS OF INTERACTIONS

Several package-product interactions can reduce the quality of packaged foods. Undesirable effects include alteration of aroma (e.g., loss of desirable or gain of undesirable), changes in package performance (e.g., loss of seal strength), or addition of potential toxicants.

2.1 Transfer of toxicants

The most well studied and documented undesirable interactions are the migration (transfer) of packaging materials from the container to the food (6). Leaching of lead from lead soldered cans and migration of plasticizers from polymers are examples. The interest in migration stems from a concern over the toxicological significance of many migrants (7).

In the case of plastic components, work over the last 3 decades has led to both an applied and theoretical understanding of migration (8). However. there are still new aspects of this interaction. For example, workers in the United Kingdom have published a series of papers investigating the migration of plasticizers during microwaving foods (9) and workers in the U.S. have studied the transfer of volatiles and non-volatiles from suseceptor materials during microwaving. The theoretical aspects of migration are analogous to those associated with permeation. Detailed reviews of migration have been published (10).

2.2 Moisture gain/loss

Water vapor permeates polymers much in the same way as do fixed gases. This transfer of moisture into or out of a container can have both quality and safety implications. Water vapor transfer can be inhibited by increasing the barrier properties of the package but usually at an increased package cost.

2.3 Oxidation

Ingress of oxygen is detrimental to many packaged foods and loss of quality can be directly linked to the rate of oxygen permeation. As with water vapor, increasing barrier properties reduces deterioration rate.

2.4 Gas loss

Loss of carbonation due to permeation from within soft drink containers represents a quality change due to a specific type of interaction. Similar problems with modified atmosphere packaging have also been noted.

2.5 Nutrient loss

Nutrients can be lost through food-package interaction by at least two mechanisms. The first is the permeation of O_2 into the packaging resulting in degradation of vitamins; loss of ascorbic acid in juices, for example (11). The second results from u.v. light transmission which likewise can destroy nutrients. Loss of riboflavin from milk in plastic containers is an example (12).

2.6 Flavor deterioration

The interaction of greatest research interest is the transfer of aromatic flavors into (i.e., sorption) or through (i.e., permeation) polymeric packaging materials (3,13). Another mechanism by which flavor can be adversely affected is by transfer of undesirable odors from polymers to foods (14).

Aromas, whether desirable or undesirable, are usually made up of a mixture of volatile organic compounds which when perceived simultaneously in the nose elicit specific responses. If this mixture is changed qualitatively or quantitatively, the perception changes. Experience with migration has shown that organic compounds differ in both their affinity for polymers and their partition between polymers and aqueous phases (15). It is not surprising then that the flavor "balance' of a food can be changed by exposure to plastics nor that some odor active components of plastics have high affinity for foods.

3 DESIRABLE EFFECTS OF INTERACTIONS

It is possible, at least conceptually, to design packaging which utilizes package-food interactions to maintain or improve quality. Examples of such desirable effects include reduction of undesirable aromas, timed release of desirable compounds (e.g., aromas, antimicrobials, antioxidants), or alteration of gas atmosphere.

3.1 Alter gas composition

The permeation phenomenon has been utilized to extend the shelf life of respiring fruits and vegetables. Respiration produces CO_2 whose concentration inside the package will depend on the rate of respiration and the rate of permeation through the package. O_2 will permeate into the package as the O_2 is used in respiration. These three factors (CO_2 and O_2 permeation, and respiration rate) will come to some equilibrium gas concentration (16). Equilibrium gas mixtures which are elevated in CO_2 and reduced in O_2 (e.g., 2-5% CO_2 and 2-10% O_2) will reduce senescence of many products (17).

3.2 Remove undesirable aromas/flavors

Packaging materials can be designed to absorb or otherwise remove undesirable flavors/odors from foods. For example, sorption of limonin, which is a bitter component of citrus oil, by polyethylene has been proposed as a method to improve orange juice flavor (18).

Removal of undesirable flavors by packaging might take a more active approach. Reduction in the limonin content of citrus juice enzymically by limonate dehydrogenase has been demonstrated (19). This enzyme could, theoretically, be covalently bound to the inner layer of a juice package. Limonate dehydrogenase would hydrolyse the limonin during storage. Other covalently bound flavor producing or flavor removing enzymes might have applications for *in situ* flavor improvement.

3.3 Inhibit oxidation

Packaging can inhibit oxidation, most simply by acting as a barrier to the ingress of O_2, but to be effective the O_2 inside the package must also be removed. This means that the package must be gas flushed or vacuumed prior to sealing. Interactive packaging systems which absorb the residual oxygen have been developed. The first and simplest types are iron containing compounds that react with O_2 to form iron oxides and thus inhibit oxidation by removing the O_2. In most cases permeable packets or sachets which contain the iron are enclosed with the food in a container. Several commercial applications of these packets have appeared in recent years (20). More recent technologies incorporate oxygen scavengers directly into films or polymers. These absorbers are usually organic redox compounds that do not migrate to foods. Current applications are as cap liners for O_2 sensitive beverages such as beer or inner film layers in flexible materials.

3.4 Inhibit microbial growth

In many cases, the shelf life of non-sterile foods can be extended by inhibiting the surface growth of microorganisms. Treatment of foods with antimicrobials has been the standard approach to inhibiting spoilage. Interactive packaging which has anti-microbial activity is an alternative approach (21). Such packaging can be divided into two types: Those in which an antimicrobial agent is incorporated into the food contact layer with the intent that it will migrate to the food surface and those which inhibit surface microorganisms without migrating. Examples of the former are sachets which emit ethanol vapors to inhibit molds in bakery products and silver metal coated zeolites in which small amounts of silver migrate to the food surface (see T. Ishitani, this volume).

The most desirable method to inhibit microbial growth might be to covalently bind a substance into the surface of a film such that it did not become part of the food yet still had antimicrobial activity. While difficult, this concept may be feasible. For example, covalent linking of an enzyme which converts food carbohydrates (i.e., sugars) into hydrogen peroxide might result in an inhibition of surface growth. Substances such as chelating agents which tie up necessary nutrients such as minerals for microbial growth may also be effective.

3.5 Addition of desirable aromas

The controlled release of desirable aromas from polymers has been commercialized by the fragrance industry but has not been utilized by the food industry. However, applications where the slow timed release of aromas from packaging either to the environment or to the food have been patented. In may be desirable in some cases to replenish or enhance aroma compounds during product storage.

4   FLAVOR INTERACTIONS AND OLFACTOMETRY

Because of its importance, the aroma of a great many foods has been studied in detail (22, 23). Several hundred volatile compounds have been identified in many products such as coffee. However, only a few of these compounds are odor active and contribute to aroma. In most cases, the two most important and fundamental questions about aroma remain unanswered: Which specific compounds are responsible for the characteristic aroma and, quantitatively, how much does each odor active compound contribute to overall aroma? Processing, packaging and storage negatively affects most aromas and maintaining fresh flavor improves most products. Knowledge of which volatile compounds are important in aroma (both desirable and undesirable) is required if flavor-package interactions are to be understood and reduced.

4.1 Olfactometry and CharmAnalysis

We have used a technique called gas chromatography-olfactometry (i.e., CharmAnalysis) to investigate the odor active compounds in foods and how they interact with packaging. Advances in mass spectrometry have made it possible to identify volatile compounds. However, identification of volatiles does not indicate which compounds, if any, contribute significantly to aroma. Without knowing which compounds contribute to aroma, it is impossible to correlate qualitative and quantitative changes in these compounds with processing or packaging variables. CharmAnalysis is a form of gas chromatography-olfactometry (GCO), which can be applied to solve this problem. This technique combines the separation ability of capillary gas chromatography, the sensitivity and selectivity of the human nose (the nose responds only to compounds which have odor activity in humans), and the data processing abilities of the microcomputer (24, 25).

In CharmAnalysis, volatiles (e.g., coffee or tea) are isolated, concentrated to a known volume, injected into the gas chromatograph, and a trained human "sniffs" the humidified effluent from the gas chromatographic column. As compounds which elicit a response (i.e., have odor activity) exit the column and are perceived, the subject presses a button on a computer to record the retention time and duration of the response. The subject indicates the character of the aroma (e.g., "burnt" "roasted" "sweet" etc.). After a complete separation, the sample is diluted by three-fold and the analysis repeated. Those volatile compounds that have only weak odor activity either because they occurred in very small amounts or they have high odor thresholds are not detected in the diluted samples. Compounds with high odor activity (i.e., low thresholds) or occur in large amounts will be detected in the dilution. A third three-fold dilution (1:9) will further eliminate odor compounds as will further dilutions. This process of dilution and reanalysis continues until no or little odor is detected, usually within 8 dilutions (8 dilutions=1:6561). The computer then sums all analyses. Those volatile compounds which produce strong odor will sum to larger "CHARM" values than compounds with weak odor activity. Compounds with no odor activity will be ignored (they produce no response and are not seen on the chromatogram). This method is dependent only on the contribution to total odor of each compound in the sample. Retention times are correlated with standards and the structures of the odor active compounds determined by mass spectrometry. Identification can be difficult, however, because the compounds sensed by the nose may occur in too low a concentration to be identified by mass spectrometry. This is not a limitation, however, because knowledge of the specific structure of a compound is not necessary to determine its importance to total aroma or the effect of processing or packaging on its odor contribution.

CharmAnalysis produces a dimensionless measure of odor intensity based on odor-detection thresholds. Sensory chromatograms are experimentally produced in which the ratio of the amount of odor-active compound eluting at a particular index to the threshold amount for that same compound is determined. The chromatogram is made plotting c against retention index, where c is a simple function of the dilution factor, d, and the number of coincident responses, n:

$$c = d^{\,n-1}$$

Peak areas measure odor intensity which depends on both the inherent odor activity of the compound __and__ the amount present. Inhibition and synergism between compounds are eliminated because the chromatographic column has separated the components. Sniffers do not have to estimate intensity of the aroma yet a quantitative measure of potency is derived. Because samples are diluted until no odor is detected, the resulting CHARM values are proportional to the amount of odor activity in the most concentrated sample and inversely proportional to the compound's threshold. The lower the threshold, the higher the CHARM value.

4.2 CharmAnalysis of orange juice

Aroma interactions have been studied by olfactometry in orange juice (26). The principal component of orange juice is limonene which comprises more than 90% of the oil. However, this doesn't indicate the importance, if any, of this compound to orange juice aroma. Figure 1 compares the flame ionization detector (FID) response (which quantitatively measures all volatile compounds present) to the Charm chromatogram (which measures only odor activity independent of amount). The FID peak at retention index (RI) 1085 is quantitatively in the highest concentration in this particular juice. GC-

MS analyses identified this compound as limonene. However, this same peak in the Charm chromatogram is small and of little significance. The largest peak in the Charm chromatogram is 1085 which was identified as linalool. Charm peaks in the RI range of 1300 to 1400 produce significant odor but are not detectable in the FID chromatogram. Table 1 lists by deceasing order of aroma activity (measured in Charm units) the 16 odor active peaks from orange juice. Limonene contributes 7 of a total of 3808 units while linalool (RI=1085) contributes 554 or 80-fold more aroma. As a total, limonene contributed only 0.2% of the aroma (26).

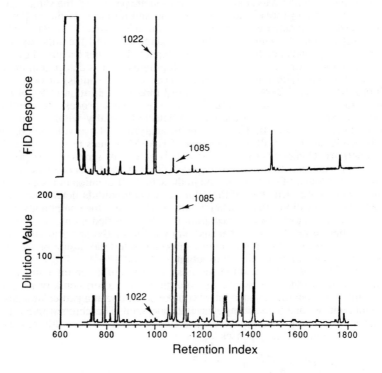

Figure 1 *Gas chromatograms of orange juice aroma detected by flame ionization (upper chromatogram) and olfactometry (lower chromatogram). Peak at retention index of 1022 was identified as limonene, peak 1085 is linalool (see reference 26).*

Once the contribution to total aroma of each compound is known the effects of packaging on individual and total aroma can be determined. For example, we exposed orange juice to both LDPE and Surlyn and determined the change in the Charm chromatogram (Table 1). Total aroma decreased by 37 and 30% for LDPE and Surlyn, respectively. While this decrease may seem large, the human nose is not able to quantitatively discriminate between aroma intensity with accuracy and these differences are borderline. That is, the effects of these polymers on aroma are generally small in comparison to the ability of humans to detect differences in aroma intensity. Individuals trained in orange juice aromas are likely to discern a difference, untrained individuals may not.

Table 1 *Retention indices and Charm values for aroma peaks from orange juice exposed to LDPE or Surlyn compared to unexposed orange juice (control) (see reference 26).*

RI	Control	LDPE	Surlyn	ldpe/con	sur/con
1237	1103	266	212	0.24	0.19
1085	554	465	250	0.84	0.45
1344	391	71	49	0.18	0.13
1358	127	194	52	1.53	0.41
1408	41	316	350	7.71	8.54
1062	19	104	106	5.47	5.58
Total Charm	3808	2390	2650	0.63	0.70

4.3 CharmAnalysis of plastic aroma

We have applied similar techniques to identification of undesirable "plastic" odors resulting from processing LDPE (27). We first identified nearly 100 volatile compounds resulting from the thermal oxidation of LDPE (28). As pointed out above, this does not indicate which if any of these compounds contribute to undesirable odor. CharmAnalysis revealed that the most odor active compounds were not normally detectable in either the FID nor GC-MS chromatograms and would likely have been overlooked (Table 2).

Exhaustive GC-MS analyses and comparison to standard compounds revealed the structures of the most active compounds. However, the structure of the compound with the highest Charm value (RI=1073; 32% of total Charm) could not be identified with confidence because of its extremely low concentration. Partial identification indicated it is an α-unsaturated carbonyl compound. The structures of the next three highest Charm compounds were confirmed as C_{7-8-9} α-unsaturated carbonyls. These 5 compounds accounted for 76% of the total aroma from thermally oxidized LDPE. Clearly, the α-unsaturated carbonyl structure in general is highly odor active and responsible for "plastic" aroma. Detection of this compound can be used as a quality control measure for plastic manufacture or an end point to determine processing parameters which minimize off odors due to polymer processing.

5 CONCLUSIONS

Food packaging is not inert and interacts directly with foods. Polymeric packaging is more interactive than most other forms. These interactions can have negative effects but the same interactions can be used to develop packaging which benefits packaged foods.

Of the several types of interactions, those dealing with aroma are the most widely investigated. In understanding how polymers affect aroma it is important to know what compounds are the most odor active and how these compounds interact with packaging. In this way, solutions to this problem can be found and improved packaging materials developed.

Acknowledgements

Portions of this paper were based on the thesis work of A. Bravo submitted to Cornell University.

References

1.   D. Gray and J. Guillet, *Macromolecules*, 1972, **5(3)**, 316.
2.   A. Roland and J. Hotchkiss, *ACS Symp. Series*, American Chemical Society,

Table 2 Retention indices, percent of total aroma, concentration, and identification of odor active compounds resulting from the thermal oxidation of LDPE. (see reference 27).

No	index	Charm %	µg/g PE	character	compound
1	1073	32.33	9	waxy	unsaturated carbonyl
2	959	18.14	<2a	herb/metallic	1-octen-3-one
3	1059	13.51	<2a	pungent	1-nonen-3-one
4	746	7.25	—b	waxy	unknown
5	856	5.40	<2a	herb/metallic	1-hepten-3-one
6	1210	2.92	—b	dusty	unknown
7	1079	2.82	97	rancid/other	nonanal
8	1175	2.26	—b	waxy	unknown
9	982	2.21	107	orange	octanal
10	1137	2.16	11	paper/carton	2-nonenal
11	655	1.72	—b	rancid/cheesy	diacetyl $CH_3-CO-CO-CH_3$
12	772	1.45	177	herbaceus	hexanal
13	1124	0.98	—b	countryside	unknown
14	814	0.59	—b	rancid/cheesy	unknown

(a) Limit of detection = 2 µg/g PE.
(b) Insufficient peak size or overlapping peaks.

Washington, DC., 1991, **473**, 149.
3. J. Landois-Garza and J. Hotchkiss, *ACS Symp. Series*, American Chemical Society, Washington, DC., 1988, **365**, 42.
4. M. Shimoda, T. Nitanda, N. Kadota, H. Ohta, K. Suetsana, and Y. Osajima, *Nippon Shokuhin Kogyo Gakkaishi*, 1984, **31**, 697.
5. C. Mannheim, J. Miltz, G. Ben-Ayrie, and R. Lavie, Programs and Abstracts, IFT National Meeting, 1984, Abs. No. 131.
6. L. Castle, A. Mercer, and J. Gilbert, *Food Addit. & Contam.*, 1991, **8(5)**, 565.
7. N. Crosby, "Food Packaging Materials: Aspects of Analysis and Migration of Contaminants", Applied Science Publishers, London, 1981.
8. S. Chang, C. Guttman, I. Sanchez, and L. Smith, *ACS Symp. Series*, American Chemical Society, Washington, DC., 1988, **365**, 106.
9. L. Castle, J. Nichol, and J. Gilbert, *Food Addit. & Contam.*, 1992, **9(4)**, 315.
10. D. Till, A. Schwope, D. Ehntholt, K. Sidman, R. Whelan, P. Schwartz, and R. Reid, *CRC Crit Rev. Toxicol.*, 1987, **18(3)**, 215.
11. G. Sadler, M. Parish, and L. Wicker, *J. Food Sci.*, 1992, **57(5)**, 1187.
12. A. Munoz, R. Ortiz, and M. Murcia, "Determination by HPLC of Changes in Riboflavin Levels in Milk and Nondairy Imitation Milk During Refrigerated Storage", Elsevier Applied Science, Essex, 1994, Vol. 49, Chapter 2, p. 203.
13. G. Halek and J. Luttmann, *ACS Symp. Series*, American Chemical Society, Washington, DC., 1991, **473**, 212.
14. G. Durst and E. Laperle, *J. Food Sci.*, 1990, **55(2)**, 522.
15. G. Strandburg, P. DeLassus, and B. Howell, *ACS Symp. Series*, American Chemical Society, Washington, DC, 1991, **473**, 133.
16. A. Exama, J. Arul, R. Lencki, L. Lee, and C. Toupin, *J. Food Sci.*, 1993, **58(6)**, 1365.
17. L. Gorris and H. Peppelenbos, *Hort. Technol.*, 1991, **2(3)**, 303.
18. M. Manlan, R. Matthews, R. Rouseff, R. Littell, M. Marshall, H. Moye, and A. Teixeira, *J. Food Sci.*, 1990, **55(2)**, 440.
19. L. Brewster, S. Hasegawa, and V. Maier, *J. Agric. Food Chem.*, 1976, **24(1)**, 21.
20. T. Klein and D. Knorr, *J. Food Sci.*, 1990, **55(3)**, 869.
21. Y. Weng and J. Hotchkiss. *Pkg. Technol. & Sci.*, 1993, **6**, 123.
22. S. Chang, C. Guttman, I. Sanchez, and L. Smith, *ACS Symp. Series*, American Chemical Society, Washington, DC., 1988, **365**, 106.
23. M. Shimoda, K. Wada, K. Shibata, and Y. Osajima, *Nippon Shokuhin Kogyo Gakkaishi J. Jap. Soc. Food Sci. Technol.*, 1985, **32(6)**, 377.
24. T. Acree and J. Barnard, "Trends in Flavour Research", Elsevier Sciences B.V., 1994, p. 211.
25. T. Acree, J. Barnard, and D. Cunningham, *Food Chem.*, 1994, **14**, 273.
26. A. Marin, T. Acree, and J. Hotchkiss, *J. Agric. Food Chem.*, 1992, **40**, 650.
27. A. Bravo, J. Hotchkiss, and T. Acree, *J. Agric. Food Chem.*, 1992, **40**, 1881.
28. A. Bravo and J. Hotchkiss, *J. Appl. Polymer. Sci.*, 1993, **47**, 1741.

Factors Affecting Permeation, Sorption, and Migration Processes in Package-product Systems

J.R. Giacin

SCHOOL OF PACKAGING, MICHIGAN STATE UNIVERSITY, EAST LANSING, MI 48824-1223, USA

1. INTRODUCTION

In addition to the advantages that plastic materials provide in food and beverage packaging, there are corresponding concerns related to product/package interactions[1]. This is a broad base topic associated with the mass transport of gases, water vapor and low molecular mass organic compounds between product, packaging material and storage environment. Mass transport in package systems encompasses a number of phenomena referred to as either permeability, sorption or migration. Permeability includes the transfer of molecules from the product to the external environment through the package, or from the storage environment through the package to the product. Sorption involves the take up of molecules contained by the product into, but not through the package, while migration is the passage of molecules originally contained by the package itself into the product. The mass transfer process provides the basis of further physicochemical activities within the package system. Such activities may induce physicochemical changes in the product, as well as physical damage of the package, or both[2]. In a package/product system, we refer to the mass transfer processes, and to the physicochemical activities associated with them, as product/package interactions. New packaging barrier polymers reaching the market are designed to minimize interactions, and have shown that they can indeed protect foods and beverages for long periods of time.

Permeation through a polymer film or sheet is a measure of the steady state transfer rate of the permeant, which is normally expressed as the permeability coefficient, P. The permeability coefficient can be described in terms of two fundamental parameters, namely, the diffusion and solubility coefficients. The diffusion coefficient D, being a measure of how rapidly penetrant molecules are advancing through the barrier, and the solubility coefficient S, describing the amount of the transferred molecule contained or dissolved in the film or slab at equilibrium conditions.

The simplest and most common relationship relating P, D and S is given by equation (1).

$$P = D \times S \tag{1}$$

Equation (1) is applicable for situations where D and S are independent of permeant concentration within the polymer material. However, when the permeation process involves highly interacting penetrant vapors with the polymer, this simple relationship may no longer be strictly valid, since time dependent diffusion processes and non-ideal (non-Fickian) diffusion may take place. In this case longer times are required to reach a steady state process[3-6].

The loss of volatile low molecular mass organic compounds from a food into polymeric packaging materials, based on a sorption mechanism, has been and continues to be the subject of considerable attention and concern. Sorption, or the uptake of volatile components by the polymeric packaging material from the food, may also result in increased permeability to other permeants, lower chemical and mechanical resistance of the packaging material, or may affect the kinetics of the migration process. The overall effect may result in the loss of aroma and flavor volatiles associated with product quality, as well as other volatile organic food components during package storage. In food product/package systems, the characterization of sorption behavior is necessary for quality control and prediction of change in product quality, as related to the loss of components associated with product shelf life. Sorption is measured as a function of sorbate concentration by a sorption equilibrium isotherm, that can be described by Henry's law or other mathematical models. For a specific value of concentration, the partition coefficient K is a practical way to describe the change in organic sorbate concentration, either in the food or packaging, from the moment that food product and packaging material are contacted, up to the moment they reach equilibrium. The partition coefficient K is defined as the ratio of concentration or solubility coefficient of a component within a fluid phase (food phase), and the concentration of the component in the polymer material, at equilibrium. The diffusion coefficient determines the dynamics of the sorption process. The larger the value of D, the shorter the time to reach equilibrium.

Since sorption and migration are essentially the same mass transport phenomenon, migration can also be described by a partition and diffusion coefficient.

The focus of this paper is to discuss selected variables affecting the behavior of permeation, sorption and migration processes, in relation with the basic parameters such as diffusion coefficient, solubility parameters, and partition coefficient.

The variables affecting permeation, sorption and migration can be grouped as follows:

    Composition variables:
        Chemical composition of the packaging material and penetrant
        Morphology of the polymer
        Concentration of the penetrant
        Presence of co-permeant
    Environmental and geometric factors
        Temperature
        Relative humidity
        Packaging geometry

While an in-depth treatment of each of the above factors is beyond the scope of this paper, selected examples are discussed to illustrate their role in the sorption and transport of organic penetrants in barrier polymers.

1.1 Chemical Composition of the Packaging Material

As expected, the solubility and diffusivity of liquids and gases in polymers are strongly dependent upon polymer molecular structure, chemical composition and polymer morphology. Properties related to solubility, such as permeability, also behave in a similar fashion. Accordingly, from solubility theory it is expected that the solubility of an organic penetrant in a polymer is related to the difference between the solubility parameter (δ) of both the penetrant and polymer. Good solubility is expected when the difference between solubility parameter values is close to a mean zero. It should be pointed out, however, that the solubility-parameter approach is useful only in the absence of strong polymer-penetrant interactions, such as hydrogen bonding.

The relationship between penetrant transfer characteristics and the basic molecular structure and chemical composition of a polymer is rather complex, and a number of factors contribute to the sorption and diffusion processes, among the most important being:

- structure regularity or chain symmetry, which can readily lead to a three-dimensional order of crystallinity. This is determined by the type of monomer(s) and the conditions of the polymerization reaction.
- cohesive-energy density, which produces strong intermolecular bonds, Van der Waals or hydrogen bonds and regular, periodic arrangement of such groups.
- chain alignment or orientation which allows laterally bonding groups to approach each other to the distance of best interaction, enhancing the tendency to form crystalline materials.
- the glass transition temperature (Tg) of the polymer, above which free vibration and rotational motion of polymer chains occur so that different conformations can be assumed.

Polymer free volume is also a function of structural regularity, orientation and cohesive energy density. The aforementioned structure-property relationships all contribute to a decrease in solubility and diffusivity, and thus permeability.

Examples of the effect of polymer molecular structure and chemical composition on the sorption equilibrium and diffusion values for acetone vapor by a series of barrier polymer films of varying functionality are shown in Table 1.

1.2 Morphology of Polymer

Solid state polymer chains can be found in a random arrangement to yield an amorphous structure or highly ordered crystalline phase. Most polymers used in packaging are semicrystalline or amorphous materials.

Morphology refers to the physical state by which amorphous and semicrystalline regions coexist and relate to each other in a polymer.

Polymer morphology depends not only on its stereochemistry but also on whether the polymer has been oriented or not, and at which conditions of temperature, strain rate and cooling temperature.

Morphology is indeed important in determining the barrier properties of semi-crystalline polymers. This is illustrated by the results of permeation studies carried

Table 1.   *Solubility and Diffusion Coefficient Values for Acetone Vapor in Barrier Films[a]*

Film Sample	Solubility (kg/kg) x 10^2	Diffusion Coefficient (cm^2/sec) x 10^{12}
MXD-6 Nylon	1.4	5.1
PET[b]	8.7	80
PET[c]	8.4	84
High Density Polyethylene	1.3	210

[a]   Values determined at 22 ± 1°C and 250 g/m^3 acetone vapor concentration.
[b]   400% orientation, with an orientation temperature of 115°C.
[c]   400% orientation, with an orientation temperature of 90°C.

out on biaxially oriented polyethylene terephthalate (PET) films of varying thermomechanical history[7].

Film samples were oriented biaxially at a strain of 350%/sec based on the initial dimension of 4 x 4 inches, which corresponded to an orientation rate of 14 inch/sec, biaxially.

The degree of orientation was 400% based on the initial dimensions. The orientation temperature was 90, 100 and 115°C, respectively. Table 2 presents values of percent crystallinity calculated from density values of the films at the respective orientation temperatures.

Table 3 summarizes the results of permeability studies carried out with ethyl acetate in PET film biaxially oriented a 90 and 115°C, respectively, and serves to illustrate the effect of thermomechanical history (% crystallinity) on the relative barrier properties of PET, for the permeation of ethyl acetate. The percent crystallinity of PET film oriented at 90°C was 22%, while the percent crystallinity of the film sample oriented at 115°C was 31%. As shown, ethyl acetate permeability values decreased by approximately four times by increasing the film orientation temperature from 90 to 115°C.

The solubility of vapors and gases in polymers is also strongly dependent on crystallinity, since solubility is usually confined to the amorphous regions. This is shown by the results of sorption studies carried out on the oriented PET film samples.

The equilibrium solubility and solubility coefficient values obtained for ethyl acetate in the PET films are presented in Table 4.

The solubility parameter of ethyl acetate is 9.1 (cal/cm^3)$^{1/2}$ and that for PET is 10.7 (cal/cm^3)$^{1/2}$. Since the polymer and penetrant are of similar polarity, penetrant/polymer interaction, resulting in the swelling of the polymer structure, is not unexpected.

Table 2.   *Density, and Mass-Fraction Crystallinity of the PET Sample Films.*

Orientation Temperature, °C	Density g/cc	Mass-Fraction Crystallinity Percent
90	1.360	24
100	1.366	30
115	1.371	33

Table 3.   *Permeability of Ethyl Acetate Through PET Film Biaxially Oriented at 90 and 115°C*

Orientation Temperature (°C)	Vapor Activity (a)	Run Temperature (°C)	P[a] Permeability Coefficient x 10^{20}	D Lag[b] Diffusion Coefficient x 10^{12}
90	0.59	30	2.6	1.8
	0.43	37	4.8	2.9
	0.21	54	15.4	11.0
115	0.59	30	0.014[c]	-
	0.21	54	3.6	5.3

[a]   Permeability Coefficient Units are kg m/m^2.s.Pa.
[b]   Diffusion Coefficient Units are cm^2/sec.
[c]   No permeation after 550 hrs. Value of P reported represents an upper bound.

Table 4.   *Solubility of Ethyl Acetate Vapor in PET Film at 37°C.*

Orientation Temperature (°C)	S_e Solubility kg Vapor/kg Polymer	S Solubility Coefficient kg/kg·Pa x 10^7
90	0.016	8
115	0.011	5

1.3 Concentration Dependence of the Transport Process

The results of studies on the permeance of limonene vapor through (i) oriented polypropylene, (ii) Saran-coated oriented polypropylene (iii) two-sided acrylic (heat seal) coated biaxially oriented polypropylene, and (iv) one-side Saran coated, one-side acrylic coated polypropylene film samples, as a function of penetrant concentration, are presented graphically in Figure 1, where permeance is plotted as function of penetrant concentration[8]. The observed concentration dependency of the permeance values may be attributed to penetrant/polymer interaction, resulting in configurational changes and alteration of polymer chain conformational mobility. Zobel[9] reported similar findings for the transport of the penetrant benzyl acetate through coextruded oriented polypropylene and Saran coated oriented polypropylene, at various penetrant concentrations.

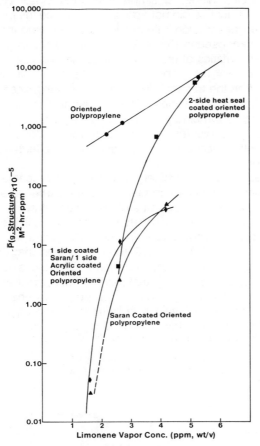

Figure 1   *The effect of limonene vapor concentration on log $\bar{P}$ for oriented polypropylene and coated oriented polypropylene structures (21-23°)*

1.4 Presence of Co-Permeant

As previously shown, organic vapors are capable of exhibiting concentration dependent mass transport processes. Therefore, the type and/or mixture of organic vapors permeating will determine the magnitude of sorption and permeation, as well as the effect of a co-permeant on penetrant permeability. The synergistic effect of a co-permeant is illustrated by the results of permeability studies carried out on a biaxially oriented polypropylene film. The degree of film orientation was 430% (machine direction) and 800% (cross machine direction), based on the initial dimensions. Binary mixtures of ethyl acetate and limonene of varying concentration were evaluated as the organic penetrants[10].

Results of permeation studies for selected ethyl acetate/limonene binary vapor mixtures are presented in Figures 2 and 3, respectively. As shown in Figure 2, (ethyl acetate a = 0.10 and limonene a = 0.18), limonene vapor had a significant effect on the transport properties of the co-permeant. A 500% increase in the permeability coefficient of ethyl acetate was obtained when compared to ethyl acetate vapor permeability alone, at similar test conditions. However, at this concentration level, ethyl acetate did not appear to influence the permeation of the limonene vapor. The transmission rate profile curve for limonene vapor in the binary mixture is superimposed in Figure 2, to provide a complete description of the transmission characteristics of the mixed vapor system.

For the ethyl acetate a = 0.1/limonene a = 0.29 binary mixture, a permeation rate 40 times greater than the transmission rate of pure ethyl acetate vapor, of an equivalent concentration, was obtained. This is illustrated in Figure 3, where the transmission profile plot of the binary mixture is presented, and compared to the transmission rate profile curve for ethyl acetate vapor alone. Again, at this concentration level, ethyl acetate did not appear to effect the permeability characteristics of limonene vapor.

For studies carried out with the binary mixture of ethyl acetate a = 0.48 and limonene a = 0.18, the individual components of the mixture were found to have a significant effect on the permeation rates of the co-penetrant.

Although sorption equilibrium values were not determined in the present study, by comparing permeability and diffusivity data it appears that the solubility coefficient of limonene is much higher (10-100 times) than that of ethyl acetate. This difference can be supported by the numerical values of the solubility parameter of the components of this system, 9.1 $(cal/cm^3)^{1/2}$ for ethyl acetate, 7.8 $(cal/cm^3)^{1/2}$ for limonene, and 8.1 $(cal/cm^3)^{1/2}$ for polypropylene. The difference between the solubility parameters values of limonene and polypropylene is less than 0.5, while the difference for ethyl acetate and polypropylene is 1.0.

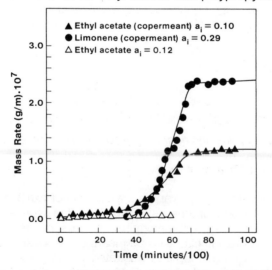

Figure 2   *Comparison of the transmission profile of the binary mixture, ethyl acetate a = 0.1/limonene a = 0.18, with the transmission profile ethyl acetate (a = 0.12) through oriented polypropylene*

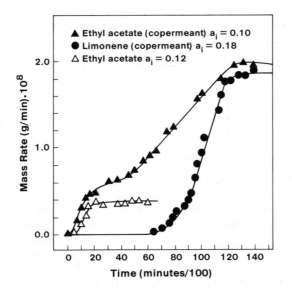

Figure 3   *Comparison of the transmission profile of the binary mixture, ethyl acetate a_i = 0.1/limonene a_i = 0.29, with the transmission profile ethyl acetate (a_i = 0.12)*

Accordingly, from solubility theory it is expected that the value of solubility for limonene in polypropylene should be higher than for ethyl acetate. This may explain the fact that only at the highest ethyl acetate activity levels is the permeability of limonene affected. These considerations are more apparent since other types of interactions, such as hydrogen bonding are not expected.

1.5 Effect of Relative Humidity

To illustrate the effect of water activity or moisture content on the barrier properties of hydrophilic polymer films, studies involving the permeability of acetone vapor through amorphous polyamide (Nylon 6I/6T) were conducted under dry conditions and in the presence of a humid environment. These experiments were conducted at 60, 75, 85 and 95°C, at a constant penetrant partial pressure value of 92 mmHg (0.29 g/l). Water activity (a_w) of the penetrant steam was maintained at 0.7 (70% RH) when measured at 23°C. Different film samples were used for dry and humid condition experiments. For an experiment, each run was maintained for a period of 8 to 14 days after attaining steady state, to ensure the system was at equilibrium[11].

A summary of the respective permeability coefficient values is presented in Table 5. As shown, sorption of water vapor resulted in an increase in permeability, as compared to dry conditions, with an increase of approximately 1.5 times being observed.

Table 5.   *Summary of Permeability Coefficient Values of Acetone Through Nylon[a] 6I/6T*

Temperature (°C)	$P \times 10^{19}$ Dry Condition	$P \times 10^{19}$ Humidified Conditions
60	3.7	4.9
75	6.5	11.2
85	9.8	17.6
95	11.8	-

[a]   Permeability expressed in $kg.m/m^2.s.Pa$.

A further illustration of the effect of water activity on the barrier properties of polymer films is presented in Figure 4, where the total quantity of ethyl acetate permeated is plotted on a function of time, for the permeability of ethyl acetate through SiO_x PET and EVAL-F films[12]. The test conditions were as follows: Temperature 22°C, 190 ppm (wt/v) concentration of ethyl acetate vapor, at 87% RH and 56% RH. Fluctuation of relative humidity was ± 2% and fluctuation of vapor concentration was ± 5%.

In over 500 hours of continuous testing, there was no measurable permeation at 56% RH. The results indicate that both test films were excellent ethyl acetate vapor barriers at 56% RH, and ambient temperature. However, as shown, at 87% RH, the EVAL-F film had a significant permeation rate of ethyl acetate vapor, while the SiO_x PET still showed no measurable rate of permeation. Table 6 summarizes the permeance values. Also presented in Table 6 are upper limit value estimations for film permeance.

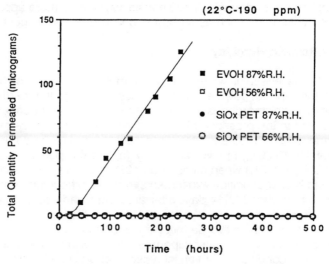

Figure 4   *Transmission rate profile curves of ethyl acetate vapor through SiO_x PET and EVOH at 22°C - 190 ppm - 87% and 56% RH*

Table 6.   *The Effect of Relative Humidity on the Permeation of Ethyl Acetate Vapor Through SiO$_x$ PET and EVOH*

RH	Sample	Permeance (kg/m^2 sec Pa) x 10^{17}
56%	SiO$_x$ PET EVAL-F	<1.1 <2.2
87%	SiO$_x$ PET EVAL-F	<2.2 840 ± 40

1.6 Packaging Geometry

In order to estimate the change in sorbate concentration associated with the sorption process, as well as that associated with migration of low molecular weight organic compounds from the package to a contact product phase, the partition coefficient (K) is required. K is defined as

$$K = \frac{C_f^*}{C_p^*} \tag{2}$$

Where C_f^* and C_p^* are the equilibrium concentrations of the sorbate in the contact phase (i.e. food product) and packaging material, respectively. In addition, the volume or mass ratio of the food phase and packaging material is also required for quantification of the equilibrium concentration for a specific product/package/sorbate system.

For geometrically simple forms, Table 7 gives the relationship between the volume of the food Vf contained by the respective packaging shapes, when the volume of the packaging material is Vp. The table compares all packages with the same thickness.

Equations (3) and (4) describe the equilibrium concentration levels in the case of migration (mC_f^*) *and sorption* (SC_f^*), as a function of initial sorbate concentration C°, and Vp/Vf.

$$W_m C_f^* = \frac{C_p^o}{K-1 + (V_f/V_p)} \tag{3}$$

$$sC_f^* = \frac{K\, C_f^o}{\dfrac{V_p}{V_f} + K} \tag{4}$$

Table 7.   Relationship Between Package Shape and Product-Package Volume Ratio

Package Shape	V_f/V_p
Rectangular	$0.150\ V_f^{1/3}$
Cubic	$0.167\ V_f^{1/3}$
Cylindrical	$0.818\ V_f^{1/3}$
Sphere	$0.207\ V_f$

REFERENCES

1.   M. Salame, *J. Plastic Film and Sheeting*, 1986, **2**, 321.
2.   B. R. Harte and J. R. Gray, "Food Product-Package Compatibility Proceedings", J. I. Gray, B. R. Harte and J. Miltz, (Eds.), Technomic Publishing Co., Inc., Lancaster, PA, 1987.
3.   E. Bagley and F. A. Long, *J. Am. Chem. Soc.*, 1958, **77**, 2172.
4.   J. Fujita, *Fortsch-Hochpolym-Forsch*, 1961, **3**, 1.
5.   J. Crank, "The Mathematics of Diffusion", Clarendon Press, Oxford, England, 1975, 2nd Edition.
6.   A. R. Berens, *Polymer*, 1977, **18**, 697.
7.   R. J. Hernandez, J. R. Giacin, K. Jayaraman and A. Shirakura, "Proceeding ANTEX '90", Dallas, Texas, May 7-11, 1990.
8.   J. R. Giacin and Hernandez, R. J., "Activities Report of the R & D Associates Proceedings of the Fall 1986 Meeting of Research and Development Associates for Military Food and Packaging Systems, Inc.", 1987, **39**, 79.
9.   M. G. R. Zobel, *Polymer Testing*, 1982, **3**, 133.
10.   T. M. Hensley, J. R. Giacin and R. J. Hernandez, "Proceedings of IAPRI 7th World Conference on Packaging", Utrecht, The Netherlands, April 14-17, 1991.
11.   S. Nagaraj, M.S. Thesis, Michigan State University, 1991.
12.   T. Sajiki and J. R. Giacin, *J. Plastic Film and Sheeting*, 1993, **9**, 97.

Safety of Food by Packaging

Z. Harmati

SWISS FEDERAL LABORATORIES FOR MATERIALS TESTING AND
RESEARCH, DEPARTMENT 'PACKAGING/PAPER', CH-9001
ST. GALLEN, SWITZERLAND

1 INTRODUCTION

Packaging is either given by nature (e.g. coconut, watermelon, egg, etc.) or manmade (e.g. bag, box, bottle, can, drum, etc.). Each product needs its own packaging. It is required by law that foodstuff must not be influenced by its packaging[1,2]. In practice interactions can not always be completely excluded. Acceptability depends mostly on the quality of materials used as packaging and on their quality of manufacturing as well.

1.1 Packaging Materials

A wide field of application is given by the utilization of various materials for packaging purposes.

1.1.1 Plastics. Compounds for the purpose of sealing beverage bottles, manufacturing of crockery (e.g. disposable cups and plates), producing films with high barrier properties by coating, coextrusion, etc..

1.1.2 Paper and Board. Materials intended to come into contact with foodstuff, such as papers for cooking, baking, and hot filtering (e.g. tea bags) etc..

1.1.3 Sheet metals . Mostly steel and aluminium for the purpose of canning fruit, vegetables, meat, etc..

1.2 Need for Testing

Any decision, which packaging shall be used for a certain food, should be preceded by specific laboratory tests. A well known procedure is the test concerning migration (Figure 1), qualitatively[3] (overall migration) and quantitatively[4] (specific migration) as well.

The question, which components may cause trouble, can often be answered by means of instrumental analytical detection methods, for instance gas chromatography and mass spectrometry. On the other hand the fact of effective interaction between a food and its packaging can often be detected by sensory testing[5].

2 DESCRIPTION OF THE TERM 'SAFETY'

What packagings have in common is to protect what they contain and to help to maintain health (active protection function). It would hardly be reasonable to put the various packaging functions in a rigid order of priority. Each function could be the most important one depending on the purpose it serves. The term 'safety' sums up a number of legislative requirements which are demanded upon packagings. In practice, a package is regarded 'safe', if it has successfully passed relevant tests to prove that it meets with all or at least with most of these requirements.

2.1 Life Cycles of Packagings

In the last few years a new dimension of safety has arisen, the 'ecological dimension'. This means that packagings have not only to satisfy physical, chemical and biological criteria during their 'lifecycle' as packagings. Once the original function has been fulfilled packagings become 'waste'. A new 'lifecycle' starts and the new function which occurs from this moment is to decay without polluting the environment (passive protection function).

2.2 Overall Safety

In other words: 'overall safety' is demanded from packagings! First they have to protect the quality of their contents (primary function during the first lifecycle) and later on they must not endanger health by polluting the environment (main function during the new lifecycle). Natural packagings satisfy both these demands. Unfortunately this can not be said for all artificial packagings.

3 INTERACTIONS, COMPATIBILITY

3.1 Product Groups

The testing of compatibility between food and packaging material represents a special discipline of scientific activity. Though the assessment of safety remains identical for any product, the testing techniques and methods are to be adapted according to the particular demands on the packaging in test. Thus, different product groups, such as food or chemical-technical products, follow a similar testing philosophy but different procedures and criteria[6].

3.1.1 Foodstuff. Food must be kept free from any risk of contamination by components from the packaging materials. According to legislation packaging must not cause any off-taste nor change any characteristics of the food contained[1-4]. This demands laboratory testing of the chemical resistance of the packagings by means of appropriate food simulants (Figure 2).

3.1.2 Chemicals. In comparison to food packagings those of chemical-technical products, so called 'dangerous goods', must prevent leakage under all circumstances. This demands again specific performance testing of such packagings which is usually carried out following compulsory international regulations[7,8] to prove their safety for multimodal transport. The compatibility must be tested with original products or by means of simulation with defined standard liquids as well.

For both mentioned product groups the interaction between contents and packaging should be kept to an absolute minimum. Neither deterioration of the content nor corrosion of the packaging should occur. Very often however the 'enemies sit inside'. Certain ingredients in canned food (e.g. tomato sauce, pineapple, orange juice, etc.) and in non-food products as well may attack and destroy their own packaging especially at 'weak points', where existing[9,10]. That's why in many cases the packaging has to be protected against its contents as well.

3.2 Experimental Approach

In Switzerland packaging materials for food have to undergo a migration check and a sensory check.

3.2.1 Testing of Migration. The testing procedure and the evaluations follow exactly the methods and prescriptions[3,4] actually given by the competent national authority. Therefore they are specific for Switzerland. At the time being they differ in some details from those of the other European countries; however harmonization with the testing requirements in the EU-Directives is planned. Figure 1 and Figure 2 show details of the Swiss way of migration testing schematically. The testing conditions mentioned in Figure 1 (time and temperature) may vary, if special practical aspects (e.g. disposable cups for hot beverages) are to be considered. For the reason of better reproducibility the original food in test is usually replaced by defined 'simulants'. Which simulant (water, acetic acid, alcohol, olive oil, pentane or ethyl acetate) substitutes which food depends on the type of the food as indicated by some examples in Figure 2.

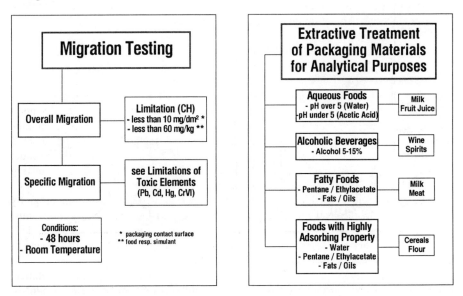

Figure 1 *Scheme for testing the migration behaviour of packagings*

Figure 2 *Examples for food substituting extraction agents*

3.2.2 Off-taste of mineral water due to migration. It is a matter of fact that substances can pass over from the packaging material into the packaged food. If this occurs off-taste may result. In case of such an unwanted phenomenon usually various reasons concerning materials and manufacturing must be taken in account. To do this and to find out which one is really responsible for the off-taste is a challenging task for the analytical chemist.

Investigations by means of solid-liquid extraction (coating compound in contact with mineral water at elevated temperature during two days) and gas chromatographic determination (GC) of the components migrating into the liquid phase brought out that not every plastic compound is suitable as coating of metal caps of glass bottles for beverages. Figure 3 shows the Total Ion Chromatogram (TIC) proving the presence of traces of at least four different organic substances in the mineral water, such as 0.3 ppm **2-hexanol** (peak no.1), 4.4 ppm **3,5,5-trimethylhexanol** (peak no.2), 1.6 ppm **nonanol** (peak no.3) and 12 ppm different **alcohols C9-C11** (peak no.4), due to water soluble impurity of the polyvinychloride (PVC) coating of the metal caps used on mineral water bottles. The identification of the found peaks in the TIC was done by means of mass spectrometry (MS).

Sensory tests carried out parallelly gave the evidence that each one of the identified substances, especially the nonanol and of course the mixture of all together, can be regarded responsible for off-taste of mineral water even in such very small trace amounts as found after a certain storing time.

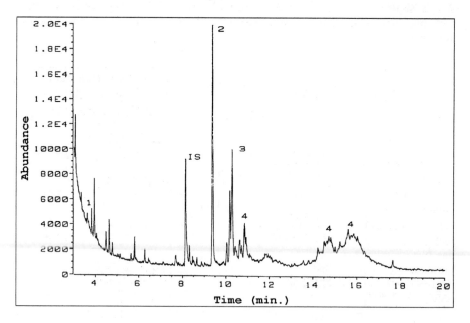

Figure 3 *Total Ion Chromatogram (TIC) of mineral water whith off-taste due to migration of impurities from the coating compound of the bottle cap basing on polyvinylchloride (PVC). The peak 'IS' signalizes benzonitrile used as 'Internal Standard'*

Comparative tests were made to check the suitability of a polyethylene (PE) based compound. For this purpose a new series of mineral water extracts was prepared and analysed in the same way. The TIC (Figure 4) shows the result. The PE-compound tested was free of migrating impurities. The only significant peak which occurs is that of benzonitrile used as Internal Standard (IS). This compound turned out far not as critical as the one based on PVC. The result could be confirmed by sensory testing as well. No significant sensory differences were detectable.

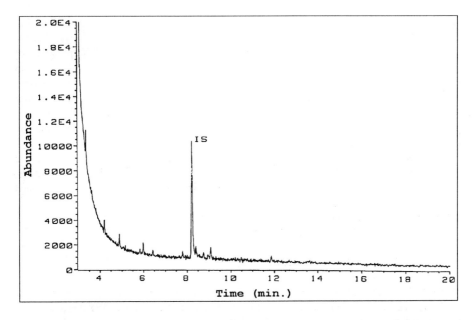

Figure 4 *TIC of mineral water as represented in Figure 3, however, without off-taste. The coating compound of the bottle cap was based on polyethylene (PE)*

3.2.3 Off-taste of biscuits due to impurity of the wrapping material. Instead of solid-liquid extraction the 'head space technique' was applied for an other investigation to figure out the reason for off-taste of biscuits being wrapped and sealed in plastic film laminated paper.

Simulation was done by means of a pouch made from the original wrapping material, sealed, and heated to 80°C for a period of 30 minutes before analysing the gas phase within the pouch by means of GC/MS. The TIC (Figure 5) proves the presence of **2-butoxyethanol**, signalled by the large peak occurring at about 6.4 minutes retention time. It is a well known solvent often used in the manufacturing of polymers.

Thus evidence was given that this solvent has not been removed entirely from the wrapping material before being used for packaging. During storage the outgasing solvent penetrated the biscuits and caused the change of taste.

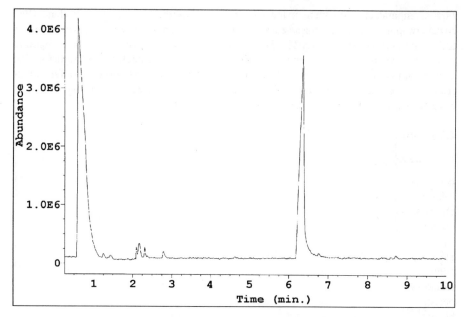

Figure 5 *TIC of the heated gas phase (80°C) in a pouch of plastic film laminated paper for biscuits taken from the pouch using 'head space technic'*

4 CRITICAL PARAMETERS

The following examples merely give an idea of the troubles which occur in practice again and again and of the serious efforts which are necessary regarding the analytical testing and the rigorous provisions to be taken as well.

4.1 Compounds

4.1.1 Impurities. Plastics and plastic coatings containing for instance residual monomers, a possible consequence of wrong manufacturing and processing parameters, might intoxicate the packaged food (e.g. vegetable oils and fats). If such organo-soluble substances enter into the food chain, deseases will occur sooner or later.

4.1.2 Chlorine. Materials containing chlorine are harmful for health and nature when they are burned without appropriate smoke filtering equipment. The state-of-the-art of the disposal technology (landfilling, burning) is very important for estmations concerning the impact on the environment. Therefore regulations exist regarding the limitation of chlorine in packaging materials[11].

Rather harmful may be chlorinated solvents (e.g. perchlorethylene) if they are not stowed away from foodstuff during storage and transport.

4.1.3 Further organic contaminants. Papers and boards are to be tested on the possible presence of antimicrobial substances[12], since their precence is not allowed in materials intended to come into contact with foodstuff. The presence of further organic substances, such as dyestuffs, fluorescent whiteners, formaldehyde, polychlorinated biphenyls, pentachlorophenol, ortho-phenyl-phenol, various solvents (Figure 4) is not only unwanted but not permitted at all or extremely limited. Some very small amounts are tolerable if they are technically not avoidable[13].

4.1.4 Inorganic contaminants. Troubles arise with a number of heavy metals, such as lead, cadmium, mercury, chromium IV, being often dispersed in all kinds of packaging materials and having a latent toxic effect on life and environment, if they enter into the food chain.

4.2 Designs

4.2.1 Design Type. Following regulations inflammable liquids (e.g. alcohol, perfumes, aroma concentrates, solvents, etc.) should never be filled in containers with a large sized or full opening, so called 'removable cover'[7,8]. Though such openings allow an easy access for operating tools (e.g. mixers, etc.), they increase the risk of spilling, leaking during transport or storage, accidents and pollution considerably.

4.2.2 Permeability. Hygroscopic food (e.g. cereals, flour, salts, sugar, honey, etc.) need moisture proof packagings. If these are plastic bags, for instance, or containers of board, a high tear resistance, a proper sealing and perfect vapour isolation is indispensable, as otherwise the food becomes wet, sticky, mouldy and perishes. Some solid substances (e.g. crystalline powders) even may dissolve or become sticky in the package during a longer period of storage or transport in countries with high air humidity.

4.2.3 Sealing. Closures of packagings are often assumed to be 'weak points', as the risk of leaking in such zones is relatively high[9]. It happens indeed that a leakage and a consequent loss of a larger amount of contents (liquids, gas, flowable solids) can not be detected in time.

Aseptically filled or bottled pharmaceutics, certain cosmetics, baby food need packagings with hermetic sealing to keep out microbes.

Tops and corks of bottles for acidic liquids offen represent a potential danger, as they are sometimes hard to open, especially for older people. It happens indeed, that the contents of such bottles spoil clothes or splash into the face and eyes.

4.2.4 Tightness. Highly valuable aromas of coffee, tea, spices, dry fruits etc. may volatize, if the packaging is not sufficiently gastight or in special cases a valve does not work correctly or the sealing is not appropriate[14,15].

5 FINAL CONCLUSION AND OUTLOOK

Taking the experiences from the different reported investigations into consideration it is evident that in future the testing of packaging materials and packagings as well should be strongly recommended in order to reach and maintain the highest possible level of safety for any product worldwide for people and environment.

Developing and marketing of new and modified products are in all countries a real need of the industry. Several testing and research institutes are busy with comparison testing of packagings and relevant products. Whenever greater projects go beyond the scope of a single company mutual work and 'Multi Client Research Projects' offer a better platform for problem solutions and realization of new packaging ideas.

At the time being the Testing and Research Institute of Switzerland, EMPA, investigates the improvement of the gas (oxygen, nitrogen, carbondioxide) and water vapor barrier property of plastic films by using the plasma coating technology.

References

1. Schweiz. Lebensmittelverordnung, Bundeskanzlei, Bern, 1991, Art. 450, p. 203.
2. Schweiz. Lebensmittelverordnung, Bundeskanzlei, Bern, 1991, Art. 459, p. 207.
3. Schweiz. Lebensmittelbuch, 'Gebrauchsgegenstände aus Kunststoff', EDMZ, Bern, 1982, Vol. 5, Chapter 48, p. 7.
4. Schweiz. Lebensmittelbuch, 'Materialien aus Papier, Karton und Pappe für Lebensmittel', EDMZ, Bern, 1994, Vol. 5, Chapter 47A 1-1.1, p. 1.
5. DIN 10 951, Sensorische Prüfverfahren - Dreiecksprüfung, 1986.
6. Z. Harmati, 'Füllgutverträglichkeit von Packstoffen', *Pack Aktuell*, Zug, 1993, Nr. 5, p. 14.
7. RID, Ordnung für die internationale Eisenbahnbeförderung gefährlicher Güter, EDMZ, Bern, 1993, Anhang V, p. 1.
8. ADR, European Agreement Concerning the International Carriage of Dangerous Goods by Road, United Nations, New York, 1992, Appendix A.5, p. 3.
9. Z. Harmati, 'Combination Packagings for Dangerous Goods and their Testing in Switzerland', *Packaging*, Watford, 1987, Vol.58, No.671, p. 27.
10. Z. Harmati, 'Korrosion, der Packstoff-Feind par excellence', *Tara*, Rapperswil, 1989, Nr. 482, p. 748.
11. Verordnung über umweltgefährdende Stoffe, Bundeskanzlei, Bern, 1991, Anhang 4.11, p. 75.
12. DIN 54 380, Prüfung auf Zusätze von antimikrobiellen Bestandteilen, 1978.
13. Empfehlung XXXVI 'Papiere, Kartons und Pappen für den Lebensmittelkontakt', Kunststoffkommission des Bundesgesundheitsamtes, Bonn, 1991, p. 73.
14. U. Ernst, 'Optimization in Heatsealing of Packaging Materials', Euro Food Pack, Proceedings of the Intern. Conf. on Food Packaging, Vienna, 1984, p. 122.
15. U. Ernst, 'Peelable Seam Systems and their Production', *Packag. Technol.& Sci.*, 1994, Vol. **7**, No. 1, p. 39.

Sensory Problems: Off-taste, Aroma Sorption, Flavour Scalping

Sensory Problems Caused by Food and Packaging Interactions: Overview and Treatment of Recent Case Studies

J. Ewender, A. Lindner-Steinert, M. Rüter, and O. Piringer

FRAUNHOFER-INSTITUT ILV, SCHRAGENHOFSTRASSE 35, D-80992 MUNICH, GERMANY

1 INTRODUCTION

Off-flavours in food packages are not only a legal but also often an economic problem when market recalls have to be made. Off-odours are due to the presence of one or several substances whose strong odours affect the food's specific aroma. These off-flavour substances might be caused by reactions in the food itself, e.g. Maillard products, or by interactions between the food and its packaging. This work deals with the solution of several kinds of off-flavour problems arising from interactions between packaging and food.

2 STATE OF THE ART

Over the years many researchers have been dealing with off-odour problems in food and packaging [1]. The Fraunhofer-ILV is specialized in dealing with off-odour problems due to interactions between food and its packaging. These interactions can be classified into three different classes of problems:

2.1 Permeation

In this case the offending substance is permeating from the environment through the packaging into the food and causes the off-flavour. A typical example for this is the recent 'anisole problem' with polyethylene (PE) films. The PE granulates involved were transported in PE bags on wooden pallets from which the halogen containing phenols and anisoles permeated through the PE bags and into the granulates. Case study I deals with this off-flavour problem in more detail.

2.2 Migration

Migration is the transfer of packaging substances into the food. In the past several problems were solved related to the migration of sensorially active components. Included in these problems e.g. are latex coatings for paperboard articles. In that case the substance phenylcyclohexene was responsible for off-odours and off-flavours in the packed product [2]. Presently under study is the problem of residual solvents coming from printing inks which have caused off-flavour problems for many years [3]. Some time ago PE was systematically tested for the presence of potential odour-active compounds [4].

2.3 Reactions between Packaging Components and Food Components

In this case of interactions chemical reactions are generally completely unexpected. Often only minimal changes in the composition of the packaging or food causes such reactions or can catalyse them. For example very small traces of diacetone alcohol produced from acetone were dehydrated to form mesityl-oxide in an ethylene-ionomer. Subsequent reactions of the mesityl-oxide with sulphur compounds in the food led to the formation of 4-mercapto-4-methylpentane-2-one [5]. This substance has a very low odour threshold level and has an intense cat urine-like off-odour which made the packed ham inedible.

The examples show that it is often very hard and expensive to find the causes of off-flavours. However, most off-odour problems have several typical characteristics in common:

- In general off-flavours occur sporadically but from different directions. Often the causes can be found in new production methods, especially for the packaging material or package.

- Presently, changes in the packaging material for ecological reasons, e.g. reduction or elimination of proven barrier layers play an important role. Often substitutions of packaging materials are carried out for environmental reasons and are not fully developed technically.

- The contamination of raw materials like e.g. plastic granulates widely distributes off-flavours to many converters. In order to solve the off-odour problem in food packaging all the packaging materials used as well as their raw materials and additives have to be considered.

- Very often you have to deal with contaminated packages the history of which is unknown. Lack of knowledge about the life cycle of a package and its packed product as well as their individual components makes the solution of the off-odour problem tremendously difficult.

- Not infrequently there is an absence of good reference samples since the complete production run shows off-flavour. In this case the solution of the off-flavour problem is significantly more difficult since there are no samples for comparison in sensory and analytical analyses.

The testing of these off-flavour problems requires a systematic approach where prior experience is necessary. In spite of the best instruments and long years of experience, numerous difficulties can occur during the solution of off-flavour problems.

- Numerous substances have the same odour. Table 1 [1] shows examples of the odour descriptions 'musty', 'painty' and 'plastic' which are used for a number of different substances. It is therefore obvious that for example the description 'musty' of an off-flavour does not give any concrete reference to the kind of off-odour substance involved.

- There is also the case where different testing panels or different testers of one panel give different descriptions for the same substance. This can be due to physiological reasons which means a weaker or stronger perception for the same stimulus intensity or the different linguistic usage by different panels.

Table 1   *Possible Chemical Compounds Related to Specific Sensory Descriptors* [1, p 11]

Descriptor	Chemical Compound
musty	2,6-Dimethyl-3-methoxypyrazine
	2-Methoxy-3-isopropylpyrazine
	2,4-Dichloroanisole
	2,6-Dichloroanisole
	2,3,6-Trichloroanisole
	2,4,6-Trichloroanisole
	2,3,4,6-Tetrachloroanisole
	Pentachloroanisole
	2,4,6-Tribromoanisole
	Geosmin
	2-Methylisoborneol
	1-Octen-3-ol
	Octa-1,3-diene
	α-Terpineol
	4,4,6-Trimethyl-1,3-dioxan
painty	Heptan-2-one
	trans,trans-Hepta-2,4-dienal
	trans-1,3-Pentadiene
	2-(2-Pentenyl)furan
plastic	Benzothiazole
	Methyl acrylate
	Methyl methacrylate
	trans-2-Nonenal
	Styrene

- Another difficulty is that the same substance shows different odours depending on its concentration. Table 2 [1] shows the variation of the flavour description of trans-2-nonenal depending on its concentration in water.

Table 2   *Variation of Taste Description of trans-2-Nonenal with Concentration in Water* [1, p 38]

Concentration [μg/l]	Taste
0,2	Plastic
0,4 - 2,0	Woody
8 - 40	Fatty
1.000	Cucumber

- Another effect which makes the solution of off-flavour problems significantly more difficult is the fact that different odour active substances overlap with one another to form an unspecified global odour. The following case studies describe such an off-flavour case.

Table 3 *Odour Descriptions and Thresholds of Potential Off-Flavour-Substances in Water*

Substance	Description	Threshold [w/w]
2,3,6-Trichloroanisole	musty, mouldy, corky	$3.0 \cdot 10^{-16}$
2,4,6-Tribromoanisole	musty	$8.0 \cdot 10^{-15}$
2-Isobutyl-3-methylpyrazine	musty	$2.0 \cdot 10^{-14}$
2,4,6-Trichloroanisole	musty, mouldy, corky	$3.0 \cdot 10^{-14}$
2,6-Dibromophenol	iodoform-like	$5.0 \cdot 10^{-13}$
1-cis-5-Octadien-3-one	metallic, musty	$1.2 \cdot 10^{-12}$
2-Methoxy-3-isopropylpyrazine	musty, potato-like	$2.0 \cdot 10^{-12}$
2,3,4,6-Tetrachloroanisole	musty, mouldy, corky	$4.0 \cdot 10^{-12}$
4-Mercapto-4-methylpentan-2-one	cat urine	$1.0 \cdot 10^{-11}$
Dimethyltrisulphide	onion, metallic	$1.0 \cdot 10^{-11}$
2-Bromophenol	phenolic, iodine-like	$2.0 \cdot 10^{-11}$
2-Methylisoborneol	musty, earthy	$2.9 \cdot 10^{-11}$
Geosmin	musty, earthy	$5.0 \cdot 10^{-11}$
6-Chloro-o-cresol	disinfectant	$8.0 \cdot 10^{-11}$
2-trans-6-cis-Nonadienal	rancid	$1.0 \cdot 10^{-10}$
1-Octen-3-one	mushroom, mouldy	$1.0 \cdot 10^{-10}$
Dimethylsulphide	sulphury	$3.3 \cdot 10^{-10}$
2,4-Dichloroanisole	musty, fruity, sweet	$4.0 \cdot 10^{-10}$
2-trans,4-trans-Decadienal	oily, fatty	$5.0 \cdot 10^{-10}$
1-Octen-3-ol	musty	$1.0 \cdot 10^{-9}$
1-Penten-3-one	sharp, fishy, oily, painty	$1.0 \cdot 10^{-9}$
2-Chorophenol	disinfectant	$2.0 \cdot 10^{-9}$
2,4-Dichlorophenol	disinfectant	$2.0 \cdot 10^{-9}$
2,6-Dichlorophenol	disinfectant	$3.0 \cdot 10^{-9}$
2,3,4,5,6-Pentachloroanisole	musty	$4.0 \cdot 10^{-9}$
Octanal	rancid fat	$5.0 \cdot 10^{-9}$
2-trans-Nonenal	fatty, cucumber	$6.0 \cdot 10^{-9}$
Decanal	rancid	$7.0 \cdot 10^{-9}$
Acetaldehyde	sweet, painty	$1.0 \cdot 10^{-8}$
Hexanal	green	$3.0 \cdot 10^{-8}$
4-Chloro-o-cresol	disinfectant	$1.2 \cdot 10^{-7}$
2-Octanone	green, fruity	$1.5 \cdot 10^{-7}$
2-Nonanone	fruity, fatty, turpentine	$1,9 \cdot 10^{-7}$
p-Cresol	pigsty, drain-like	$2,0 \cdot 10^{-7}$
4-Chlorophenol	disinfectant	$2.5 \cdot 10^{-7}$
2-Nonanol	musty, stale	$2.8 \cdot 10^{-7}$
2-Heptanol	rancid coconut	$4.1 \cdot 10^{-7}$
2-Heptanone	spicy, rancid almonds	$6.5 \cdot 10^{-7}$
2-trans-Octen-1-ol	fatty	$8.4 \cdot 10^{-7}$
2,4,6-Trichlorophenol	disinfectant	$1.0 \cdot 10^{-6}$
Phenol	medicinal, phenolic	$1.0 \cdot 10^{-6}$
Pyrazine	sweet, floral	$3.0 \cdot 10^{-4}$

- One of the biggest difficulties in these tests is the very different and sometimes very low odour thresholds of these substances. Table 3 shows a series of off-odour relevant substances with their odour thresholds over water as well as their corresponding odour description (data assembled from reference 1). Pyrazines for example, have odour thresholds depending on the structure in a range of 10^{10}. 2,3,6-tri-chloroanisole has an odour threshold of 0.3 pg/l and is therefore, depending on the matrix, near the analytical detection limit. How the position of the halogen atom can affect the odour threshold is shown by comparing the thresholds of 2-chlorophenol to 4-chlorophenol with 2 µg/l and 250 µg/l.

- The above mentioned points show that a trained sensory panel is indispensable for treating such off-flavour problems. Its members should be familiar with typical off-odours as well as with their standard descriptors. They should also have experience with GC sniffing. Only by using a sensory panel that gives precise odour descriptions can off-flavour problems be solved in combination with modern instrumental analysis methods.

The practical procedures and the step-by-step solutions of such off-flavour cases are illustrated in the following four topical case studies.

3 CASE STUDIES

3.1 Off-Odours Caused by Halogenated Phenols and Anisoles

Due to their extremely low sensory threshold levels chloro- and bromoanisoles have been clearly identified for years as the causes of off-odours [6]. The following case shows how new variations and combinations of unpredictable events can lead to a problem whose solution requires a great amount of work.

In the past year there was a series of off-odour problems in packed food over a relatively short time period occurring in different types of packages and materials that were related to one another through the use of a polymer granulate. Even though all problems had different natures all of the samples studied had several similar characteristics: 1. The description of the off-odour matched that for above mentioned trichloro- (TCA) as well as tribromoanisole (TBA). 2. In all cases the packaging contained some polyethylene. 3. The number of affected samples were relatively small which means they came from a relatively small batch. Together with the seemingly simultaneous occurrence these characteristics point to a common cause. Since there was no technical reason to assume formation of TCA or TBA in PE, the study was directed to the PE granulate because it was the simplest matrix.

The discovery of TCA led to the suspicion that the granulate had become contaminated in some way. However, the occurrence of the contamination in granulates from different manufacturers could not be explained. The TCA and TBA found in PE coated cartons came from the PE and not from the paperboard. Finally, these substances were found in food that was packed in pouches containing PE.

The causes of these problems could eventually be traced back to the presence of halogenated phenols in several of the wooden pallets used for transport. One difficulty in solving the case was finding wood samples contaminated with such phenols since in international transportation the wooden pallets are stored for only a short time in one place and no reserves exist. The testing of wood samples with obvious off-odours and those without gave a clear correlation with an analytical analysis of the wood extracts using gas chromatographic separation with ECD detection. The correlation is shown in Figure 1a and 1b. The extract was analyzed underivatized so that the 2,4,6-tribromophenol peak had some tailing on the nonpolar column. The identifications of the compounds were confirmed with the help of a high resolution mass spectrometer using the molecule ions (Figure 2).

The formation of TCA and TBA from the corresponding halogenated phenols by microorganisms is known [6] and likewise the use of such phenols as wood preservatives. If by chance a bag containing PE granulate lies on a wood pallet containing halogenated phenols it is possible that contamination of the PE layer adjacent to the wood occurs by diffusion of the anisole. Given the tremendously low threshold levels on one hand and the low thickness of the PE layers needed for food packaging on the other, it would be possible for example that several thousand packages are affected from a bag containing 25 kg contaminated granulate. With the high sensitivity of today's mass spectrometers trace amounts down to approximately 50 ppt (ng/kg) of TCA or TBA can be detected in the affected packaging. For several months we have seen no further cases of this type of off-odour problem.

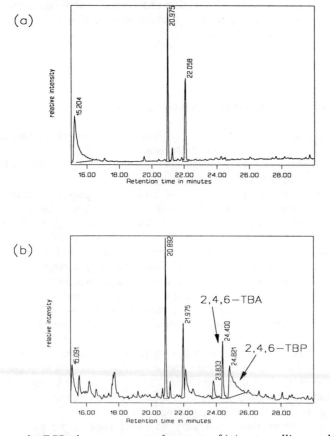

Figure 1 *ECD chromatograms of extracts of (a) not smelling and (b) smelling wooden pallets*

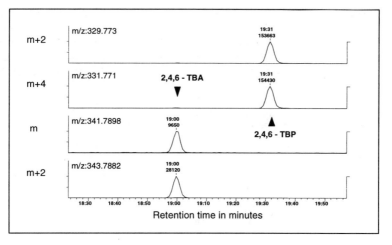

Figure 2 *High resolution mass spectrometer chromatogram of tribromoanisole (TBA) and tribromophenol (TBP)*

3.2 Off-Odour Formation by Pyrazine in Printed Cartons

A case also occurring in the past year involved a printed carton intended for food packaging use that had an external lacquer coating over the printing. The offending odour was described as musty and was similar to that of the above discussed anisoles.

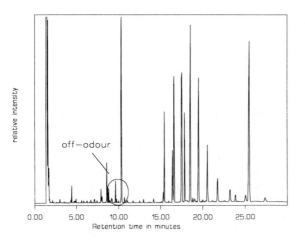

Figure 3 *FID chromatogram of an extract of bad laquer*

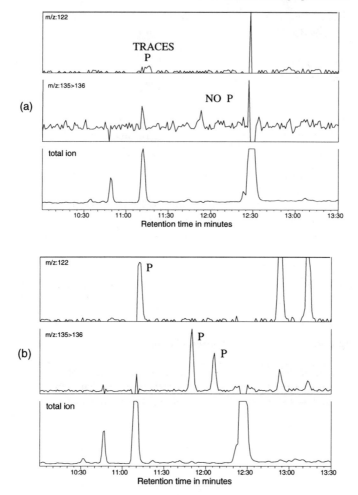

Figure 4 *Low resolution SIM chromatogram (Selected Ion Monitoring) of extracts of (a) good and (b) bad carton sample*

However, halogenated anisoles could not be found in either the carton samples or in the gas chromatographic (GC) analysis of organic solvent extracts of the odourless and off-odour containing lacquers of the same type on a capillary column using FID detection. The two lacquers had practically identical chromatograms as shown in Figure 3. During sniffing of the separated extract components at the end of the column the characteristic off-odour could be perceived in the offending lacquer sample in the circled region indicated on the chromatogram. By coupling the GC with a low resolution ion trap mass spectrometer as detector, the components designated as P in the off-odour containing carton sample in Figure 4 could be identified as pyrazines. Even though trimethyl-pyrazine was present in the good and bad lacquer samples in different concentrations, two isomers of ethyl-dimethyl-pyrazine could only be detected in the bad sample. The retention times of the pyrazines lay in the region of the characteristic off-odour during sniffing of the bad sample.

No definite conclusions could be reached with regard to the exact formation pathway of the pyrazines in the lacquer due to the limited funds available for the project. It is sure, however, that fresh lacquer is free of pyrazines and these are formed during storage in an old lacquer sample. The ammonia contained in the lacquer plays a key role in this whereby it possibly leads to amino-carbonyl compounds and by a condensation reaction to chemically stable pyrazines.

3.3 Off-Odour Formation by Superposition of Substances from Different Classes of Compounds.

The description of the off-odour from a food with a complaint ranged from tarry, mineral oil like, phenolic to musty. The packaging used was composed of a complex laminate of packaging materials. An investigation of all packaging materials pointed to the PE layer used as the off-odour carrier in the problem samples. A comparison of chromatograms of extracts from good and bad samples of the PE film showed similar patterns but with different intensities (Figure 5). From this it could be assumed that the difference between the two samples was possibly due only to a quantitative difference. The descriptions of the odour impressions perceived during sniffing the compounds eluting from the column in the region of the "mountain" shown in Figure 5b are listed in Table 4. The compounds marked by * were present only in the problem samples. Due to the above mentioned sensory impressions of "phenolic" and "musty" specific investigations for phenol

Table 4   *GC-Sniffing - Results of "Old" PE-Film*

Time	Odour
3.00	grassy, green, musty
3.20	chemical, stinking, musty
4.40	green
6.40	green, metallic
7.00	strong geranium
8.30 *	car tyre, burnt, tar - strong
9.40	chemical, carrots
10.00	carrots, musty, old
10.10	strong green, carrots, musty, old
10.40 *	exhausted, burnt, rubber
11.50 *	stinking, lavatory
12.00	pungent, urine
13.00	metallic
13.30 *	burnt, musty, lightly chemical
16.40	burnt potatoes
17.10	rancid, sweaty, chemical
19.40 *	petrol, adhesive
22.10 - 22.40	soapy, perfumed - strong
23.30	lightly burnt
24.30	strong vanilla

* = odours not present in the "new" PE-film

and halogenated anisoles were made. These substances were present however in very low concentrations, e.g. TBA near the detection limit. In addition the retention times of these substances were outside of the "mountain" shown in Figure 5b. With the help of mass spectrometry it could be shown that the main components of the "mountain" were different odour-active aromatic compounds.

The results of the investigation showed the off-odour in this case to be a superposition of various odour compounds with very different threshold levels. The source of these off-odours possibly coming from contamination introduced in a certain phase of the conversion, could not be determined due to the lack of samples from the different production steps. This case plainly shows how important it is to have sufficient sample material in determining the cause of the off-odour while at the same time how difficult it is to save samples from every step of a manufacturing process.

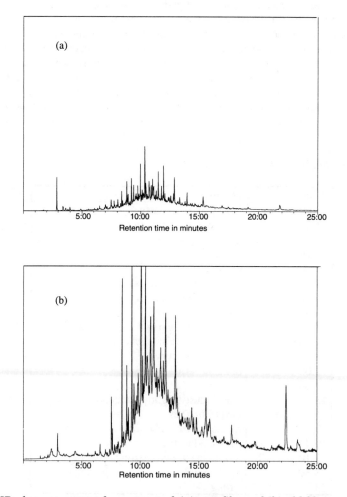

Figure 5 *FID chromatogram of an extract of (a) new film and (b) old film*

3.4 Off-Flavour Formation through Entrainment of Cleaning Agents

The importance of returnable package systems is increasing because of ecological considerations. An important step in the reuse of containers is their cleaning before refilling. A typical example is the well proven system of packaging mineral water in glass bottles. In the following case it is shown how a completely sensory neutral package like glass bottles can experience problems particularly when they are carrying sensitive products like drinking water. In a complaint with a mineral oil like off-flavour in mineral water the PE closures used were suspected at first. Upon storing a large number of these closures in water followed by extraction and analysis, no agreement was found between the closures and the water with the off-flavour. Sensory analysis of the water in contact with the closures did not show any agreement as well.

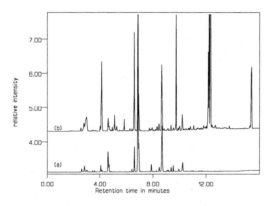

Figure 6 *Chromatogram of (a) mineral water and (b) grease*

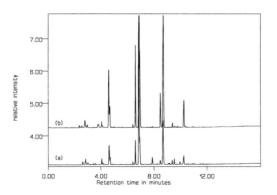

Figure 7 *Chromatogram of (a) mineral water and (b) rinsing water*

In the next step, samples were taken from all of the cleaning steps that the returnable bottles must undergo and extracts of them were compared to the water with the off-flavour. In addition all cleaning agents as well as samples of the oil and grease lubricants used for the pumps and other equipment were taken and analysed sensorically and instrumentally. It was finally shown that one of the greases used was responsible for the off-flavour (Figure 6). This grease came in contact with the bottles in the alkaline washing solution and was not completely removed in the last rinsing. The chromatogram of the rinsing water was exactly the same as that of the mineral water (Figure 7).

This case underscores the importance of sensory influences at all steps involving possible interactions between the food and its contact surface with the environment.

The above examples taken from routine experiences make clear the complexity of such off-flavour problems, as already discussed in the introduction, which are due to interactions between food and its packaging.

Acnowledgements

The authors wish to thank Ms M.Fuchs and Ms M. Zou for the preparation and extraction, Mr. L. Gruber for the MS analysis and the sensory panel for the sniffing of all the samples.

References

1. M. J. Saxby; "Food Paints and Off-Flavours", Blackie Academic & Professional, Glasgow, 1993.
2. J. Koszinowski, H. Müller, O. Piringer; *Coating*, 1980,**13**, 310.
3. M. Rüter; *Verp.-Rundschau*, 1992, **43** (8), 53.
4. J. Koszinowski, O. Piringer *Dtsch. Lebensm.-Rundschau*, 1983, **79** (6), 179.
5. R. Franz, S. Kluge, A. Lindner, O. Piringer; *Packaging Technology and Science*, 1990, **3**, 89.
6. F.B. Whitfield, T.H. Ly Nguyen, J.H. Last; *J. Sci. Food Agric.*, 1991, **54**, 595.

Permeation of Flavour Compounds Across Conventional as well as Biodegradable Polymer Films

R. Franz

FRAUNHOFER-INSTITUT FÜR LEBENSMITTELTECHNOLOGIE UND VERPACKUNG, SCHRAGENHOFSTRASSE 35, D-80992 MÜNCHEN, GERMANY

1 INTRODUCTION

The growing importance of polymeric packaging films with high protection properties against permeation of organic compounds such as food flavours or volatiles from the environment requires a better definition and understanding of what is often referred to as high barrier polymers. So far, high barrier polymers are generally defined in terms of their oxygen or other common gases barrier properties. However, water vapour and other organic substances do not necessarily agree with the permeability of gases and may therefore need separate consideration.

Whereas extensive studies have been conducted on the permeability of gases and water vapour through plastic packaging films, permeability data on flavours and other organic compounds are very limited[1,2]. One of the reasons for this obvious lack of data seems to be the non-availability of a corresponding standard test procedure which should work with high sensivity at test permeant feed vapour pressures sufficiently low to simulate properly practical packaging conditions.

Recently, a permeation test method which combines BARRER's classical test procedure for permanent gases with the adsorption and gas chromatographic determination of penetrated test permeants has been presented[3]. The method, which was found to be very sensitive (allowing mesurement of permeation rates of 1 μg permeant per square metre and day), has been validated using limonene as a model permeant and biaxially oriented polypropylene (BOPP), acrylic coated BOPP and polyethylene terephthalate (PET) as test films.

The aim of the work presented in this paper is *(i)* to review some of the permeation data of the previous work[3] and to draw practical conclusions from further evaluation of these permeation data and *(ii)* to demonstrate the broader applicability of the recently developed permeation test method employing a cocktail of seven model flavour compounds as well as a well-known off-odour principle, 2,4,6-trichloroanisole (TCA)[4], to a number of test films including both commercially available as well as experimental (biodegradable) ones.

2 MATERIALS AND METHODS

2.1 Test Films

Investigated test films are listed and specified in Table 1. Films **A** to **F** are commercially available packaging films; films **G** to **I** were experimental test films being prepared from solutions of the corresponding raw materials. Films **A** to **I** were used for model flavour permeation tests. For TCA permeation tests, structures **B** to **E**, however of different film thickness, were used: **B'** (30 µm), **C'** (32 µm), **D'** (32 µm) and **E'** (40 µm).

Table 1   *Test Films used for Permeation Measurements*

No.	Structure (thickness in µm)	Abbreviation
A	High density polyethylene (49)	HDPE
B	Biaxially oriented polypropylene (47)	BOPP
C	BOPP with acrylic coating on both sides (50)	Ac/BOPP/Ac
D	BOPP with acrylic and polyvinylidene chloride coating (50)	Ac/BOPP/PVCD
E	BOPP laminated with metallized BOPP (45)	BOPP/met. BOPP
F	BOPP laminated with polyvinylalcohol coated BOPP (44)	BOPP/PVOH/BOPP
G	Polycaprolactone (63)	PCL
H	Cellulose acetate (50)	CA
I	Polyhydroxybutyrate (BIOPOL) (60)	PHB

2.2 Test Permeants and Permeation Measurements

The seven components of the cocktail of model flavour compounds (Table 2) were supplied by Drom, Germany. 2,4,6-Trichloroanisole (TCA) was purchased from Aldrich. Other materials which were used have been described recently[3]. A mixture of the seven model flavours (relative amounts see Table 2) was dissolved in polyethylene glycol (PEG) 400 at a concentration of 5 % (w/w) to yield model flavour feed vapour pressures as given in Table 2. A 1.25 % (w/w) TCA solution in PEG 400 yielded a TCA feed vapour pressure of $1.77 \cdot 10^{-6}$ bar. For permeation measurements 40 grams of PEG 400 solution were applied to the permeation test cell to establish constant feed vapour pressures. Feed vapour pressure determinations of test permeants, permeation measurements, gas chromatographic determination of permeants and data evaluation were carried out as described recently[3].

Table 2   *Cocktail of Seven Model Flavours used for Permeation Measurements*

No.	Chemical name	Mol weight	Relative amount in the mixture of 7 model flavours	Vapour pressure over the 5% solution in PEG 400 $[10^{-6}$ bar]
1	Isoamyl acetate	130	0.05	2.9
2	Limonene	136	0.05	2.3
3	cis-3-Hexenol	100	1	3.0
4	Linalyl acetate	196	5	4.7
5	Menthol	156	10	3.9
6	Citronellol	156	10	1.1
7	Diphenyl oxide	170	10	1.0

3 RESULTS AND DISCUSSION

3.1 Diffusion and Solubility Coefficients (D and S) as Control Parameters for the Permeation of Limonene across an acrylic Layer and a PET Film

Table 3 shows a comparison of measured and calculated permeation data of limonene obtained for a BOPP, Ac/BOPP/Ac and PET film[3]. When comparing only the **P**-values as the overall effect of **D** and **S**, Ac and PET exhibit barrier properties four logarithmic orders of magnitude better than BOPP. A comparison of the kinetic (**D**) and thermodynamic (**S**) control parameters of permeation, however, indicates that the same excellent barrier behaviour is a result of two different mechanisms. Whereas the extremely low **D**-value of the Ac material indicates a purely diffusion controlled barrier principle (low **D**, high **S**) for the acrylic layer, the pemeation across PET must be considered to be a result of both a diffusion and solubility controlled process. However, compared to Ac, PET provides a better **S** controlled barrier. As a practical consequence, when having the same permeability for two different packaging films a solubility controlled permeation mechanism will be more favourable due to lower flavour absorption (flavour scalping).

Table 3 *Comparison of Three Different Polymer Materials with Respect to their Permeation Data obtained for Limonene*

Material	*Permeability coefficient* $\mathbf{P}$ [10^{-12} g/cm · s · bar]	*Solubility coefficient* $\mathbf{S}$ [g/cm^3 · bar]	*Diffusion constant* $\mathbf{D}$ [10^{-12} cm^2/s]
BOPP	156	42	3.75^b (2.06^c)
Ac	0.03^a	14^b	0.002^c
PET	0.01	0.13	0.07^b (0.06^c)

^{a} Calculated from $\mathbf{P}_{BOPP}$ and $\mathbf{P}_{Ac/BOPP/Ac}$; ^{b} Calc. from eq. $\mathbf{P} = \mathbf{D} \times \mathbf{S}$; ^{c} Calc. from the lag time t_L

3.2 Permeation of the Model Flavour Cocktail

The cocktail of the 7 model flavours was chosen from the viewpoint of applying a range of different chemical structures to the permeation test. The 7 model flavours cover a mol. weight range from 100 to 200 and represent chemical structures with a wide range of functional groups and polarities. The intention behind this selection was to use permeation data obtained from this cocktail for the prediction of the permeation behaviour of any other flavour compound. Before starting the permeation measurements the feed vapour pressures of all 7 test permeants were adjusted to the 10^{-6} bar range by mixing them inversely proportionally to their vapour pressures followed by dilution of the mixture in PEG 400 (Table 2). In this way, vapour concentrations were established without any risk of polymer swelling thus approaching real food conditions. For comparison reasons it is interesting to mention that most of the permeation studies reported in the literature apply test permeant vapour concentrations two to four logarithmic orders higher[3].

The flavour cocktail was used to test all film structures given in Table 1. Depending on the barrier properties of the test film the pemeation tests were run up to a maximum time of approx. 1600 hours. Figure 1 shows four examples of permeation curves obtained in this way. Table 4 lists the whole set of permeability coefficients calculated from the measured permeation curves.

When comparing the **P**-values given in Table 4, on a first view one can categorize on average the test films in three to four classes: *(i)* HDPE and PCL having absolutely no barrier properties, *(ii)* BOPP with very limited barrier capacity, *(iii)* Ac and PVDC coated as well as metallized BOPP films and on the other hand CA and PHB films with very high barrier functionalities and *(iv)* PVOH coated BOPP with throughout all of the 7 model flavours excellent flavour protection. In the latter case, all **P**-values were below the detection limit of the procedure. On a second, more in depth view of Table 4 it can be recognized that the test films exhibit specific permeation patterns. For example, when comparing the ratio of **P**-values for limonene (**2**), hexenol (**3**) and diphenyl oxide (**7**) then the BOPP (**B**) and the metallized BOPP (**E**) films show comparable ratios but differing completely from the ratios obtained for the Ac (**C**) and PVDC (**D**) coated films: **P**-ratio **2:3:7** = 130:73:590 (**B**) and 9.1:6.6:50 (**E**), however 7.3:4.2:17 (**C**) and 12:5.2:≤5.8 (**D**).   In other words: for film structures **C**, **D** and **E** which have very similar barrier

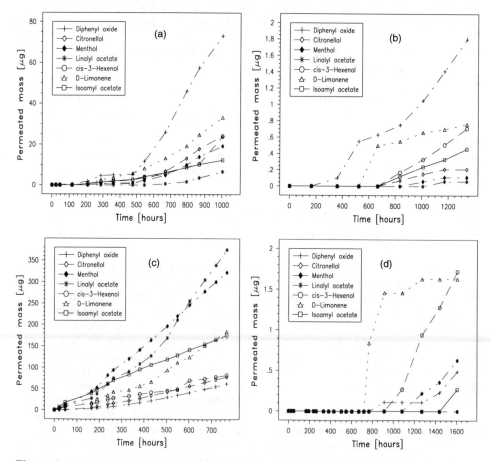

Figure 1 *Comparison of the permeation of the model flavour cocktail through different packaging films: (a) BOPP, (b) BOPP/met. BOPP, (c) PCL and (d) PHB*

Table 4   *Permeability Coefficients* **P** $[10^{-12}$ g/cm · s · bar] *of the Seven Model Flavours Measured for the Film Structures* **A** *to* **I**

Test			Model	flavour	no.		
film	**1**	**2**	**3**	**4**	**5**	**6**	**7**
A	800	4500	300	1700	2300	2100	2400
B	38	130	73	13	44	20	590
C	5.4	7.3	4.2	≤ 1.2	≤1.5	≤ 5.3	17
D	≤ 2.0	12	5.2	≤ 1.2	≤ 1.5	≤ 5.3	≤ 5.8
E	4.9	9.1	6.6	≤ 1.1	≤ 1.3	≤ 4.7	50
F	≤ 1.8	≤2.2	≤ 1.7	≤ 1.1	≤ 1.3	≤ 4.6	≤ 5.1
G	1210	3950	502	2470	2640	2270	1780
H	14	48	29	≤ 1.5	≤ 1.8	≤ 5.7	7.8
I	6.0	2.2	5.9	3.7	3.3	8.0	8.8

properties for model flavours **1** to **6** a significant differentiation with respect to the barrier against diphenyl oxide (**7**) can be found: P-ratio **C:D:E** = 17:≤5.8:50. It is assumed that this barrier selectivity is due to the aromatic ring structure of compound (**7**). For the experimental films **H** and **I** a similar discussion can be made.

3.3 Permeation of the Off-Odour Trichloroanisole (TCA)

Due to a number of TCA related off-odour complaints in packed food products occurring recently, as another test permeant TCA was investigated in a comparative study of the four test films **B'** to **E'**. The intention was to find an appropriate barrier film against penetration from the environment into the packed food. Another interesting aspect was to get additional confirmation of the assumption of the barrier selectivity towards aromatic ring structures made under 3.2. The applied TCA feed vapour pressure of $1.77 · 10^{-6}$ bar was of course very much higher than any real life situation, however, from a polymer swelling point of view it was low enough, thus allowing measurement of permeation curves in a reasonable time and at a reasonable extend of analytical expenditure.

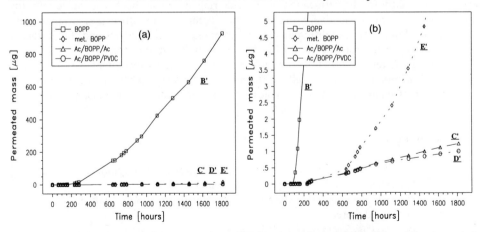

Figure 2   *Comparison of the permeation of TCA through test films* **B'** *to* **E'** *where (a) shows the full view and (b) a y-axis zoom of the permeation curves*

Table 5 *Trichloroanisole Permeability Coefficients* **P** *Measured for the Test Films* **B'** *to* **E'**

Test film	**B'**	**C'**	**D'**	**E'**
P $[10^{-12}$ g/cm · s · bar]	1600	4.2	3.5	37

The permeation curves obtained for the four test films are compared in Figure 2. The full view (a) demonstrates the dramatic increase in penetration protection when applying films **C'** to **E'** compared with **B'**. The zoomed view (b), however, reveals again the same barrier selectivity towards an aromatic ring system as already was observed for the analogous film structures **C** to **E** towards diphenyl oxide (3.2). To give a quantitative picture of this effect, the calculated permeability coefficients are presented in Table 5.

4 SUMMARY AND CONCLUSIONS

Acrylic coating and PET were found to have comparable barrier properties (**P** values) against limonene, but were four logarithmic orders of magnitude better than BOPP. From **D** and **S** values, the underlying permeation mechanisms, however, were concluded to be different. The data indicate a pure diffusion control in case of the Ac coating but for PET a permeation controlled both by **D** and **S**. It is assumed that these findings can be transferred to other non-polar flavour compounds. From a practical point of view, a solubility controlled process is the more favourable one due to lower flavour scalping.

Permeation tests on a series of commercially available BOPP films (without and with barrier layers) and experimental (made from biodegradable polymers) test films using a cocktail of 7 model flavours demonstrated the broader applicability of the permeation test method presented recently[3]. Ac and PVDC coatings as well as metallization were found to improve the barrier properties of BOPP films dramatically and comparably with biodegradeable test films made from CA and PHB. PVOH coating turned out to be the most efficient barrier of all measured films. Permeation of the off-odour principle TCA was comparatively studied on four of the BOPP film structures confirming the results obtained for the flavour cocktail. Test films were found to exhibit specific permeability patterns for the flavour cocktail, a phenomenon which could be observed most significantly for the permeation of aromatic ring systems. This barrier selectivity for aromates (meaning a relatively higher barrier compared to other non-aromatic test permeants) allows a further differentiation of otherwise similar barriers and, for instance, is in increasing order from metallized BOPP over Ac to PVDC coating.

Taking the results from this study into account it is highly recommended to consider the use of a mixture of chemically different test permeants when thinking about the introduction of standardized permeation tests for flavour or other organic compounds.

References

1. J.H. Hotchkiss (ed.), 'Food and Packaging Interactions', ACS Symposium Series 365. American Chemical Society, Washington DC, 1988.
2. K.J. Liu, J.R. Giacin and R.J. Hernandez, *Packag.Technol. Sci.*, 1988, **1**, 57 and references cited therein.
3. R. Franz, *Packag. Technol. Sci.*, 1993, **6**, 91.
4. F.B. Whitfield, T.H. Ly Nguyen and J.H. Last, *J. Sci. Food Agric.* 1991, **54**, 595.

Predictive Models for Estimation of Flavor Sorption by Packaging Polymers

J. S. Paik

DEPARTMENT OF FOOD SCIENCE, UNIVERSITY OF DELAWARE, 226 ALISON
HALL, NEWARK, DE 19716, USA

1 INTRODUCTION

For successful food package design, packaging polymers should sorb a minimum amount of the critical food flavorants while meeting all other performance requirements. Over the past few years, a number of investigators have studied the behavior of different packaging polymers in sorption of flavor compounds [1-5]. These studies show that equilibrium sorption values between polymers and flavor compounds depend on the types of polymers.

The variables that determine flavor sorption include: temperature; flavor concentration; chemical structure of flavors and packaging polymers; and polymer conformations[6]. The crystalline fraction in a polymer strongly influences the effect of polymer conformation on flavor sorption. At fixed values of storage temperature and flavor concentration in the food phase, chemical structure and polymer conformation determine the value of equilibrium flavor sorption. The sorption of a penetrant in a rubbery amorphous polymer is governed by the same factors as the solubility of compounds in organic liquids [7,8]. Diffusivity is a kinetic parameter that determines the rate of attaining equilibrium solubility but does not affect solubility *per se*.

Solubility of a given mixture is determined by its chemical make-up . Therefore, it should be possible to predict the sorption behavior in an amorphous fraction of a semicrystalline polymer using the theories of solutions. The relevant thermodynamic parameter which indicates the potential for mixing is the Gibbs free energy of mixing (ΔG_m), the free energy difference between the pure components and the mixture (Equation 1).

$$\Delta G_m = \Delta H_m - T \Delta S_m \tag{1}$$

where ΔH_m = enthalpy of mixing,
ΔS_m = entropy of mixing,
and T = temperature, K

Components are more likely to mix when the value of ΔG_m is smaller or more negative. Therefore, decreasing the enthalpy of mixing (ΔH_m) or increasing entropy of mixing (ΔS_m) will increase solution or sorption at constant temperature.

2 REGULAR SOLUTION THEORY

Van Laar[9] initially investigated the change in enthalpy during mixing of two components. Based on Van der Waals' work, Van Laar and Lorenz[10] obtained the following expression (Equation 2):

$$\Delta H_m = \{ x_1 x_2 v_1 v_2 \,/\, (x_1 v_1 + x_2 v_2) \} \, [\, (a_1/v_1)^{1/2} - (a_2/v_2)^{1/2}]^2 \qquad (2)$$

$$
\begin{aligned}
\text{where } x &= \text{mole fraction,} \\
v &= \text{molar volume,} \\
a &= \text{Van der Waals constant,}
\end{aligned}
$$

and subscripts indicate components 1 and 2.

Scatchard[11] and Hildebrand[7] developed the theory of solution by further refining Equation 2. Scatchard replaced the constant **a** in the van Laar equation (Equation 2) with $v \Delta E_v$ and defined the ration of energy of vaporization (ΔE_v) to molar volume (v) as the cohesive energy density (C) (Equation 3).

$$C = \Delta E_v / v \qquad (3)$$

Hildebrand defined the square root of cohesive energy density (C_{11}) as the solubility parameter (d) and the difference was identified as the measure of solubility characteristics of solutes and solvents.

$$\delta = C_{11}{}^{1/2} = (\Delta E_{v11} / v_{11})^{1/2} \qquad (4)$$

Enthalpy of mixing in a binary system is expressed by Scatchard-Hildebrand equation[12] as,

$$\Delta H_m = \phi_1 \phi_2 \, (x_1 v_1 + x_2 v_2) \, (\delta_1 - \delta_2)^2 \qquad (5)$$

$$
\begin{aligned}
\text{where } \quad & \phi_1 = x_1 v_1/(x_1 v_1 + x_2 v_2) & (6) \\
\text{and} \quad & \phi_2 = x_2 v_2/(x_1 v_1 + x_2 v_2) & (7)
\end{aligned}
$$

1.   The development of the Scatchard-Hildebrand equation required two major assumptions: Changes in entropy and volume upon mixing are negligible ($\Delta S_m = 0$, $\Delta V_m = 0$). Applying this assumption to Equation 1 sets the Gibbs free energy of mixing (ΔG_m) equal to enthalpy of mixing (ΔH_m). Therefore, the square of the difference between solubility parameters ($\delta_1 - \delta_2$)^2 determines the solubility. According to Scatchard-Hildebrand equation, the solubility of two components increases if their cohesive energy densities are similar, corresponding to a decrease in ΔH_m.

2.   The equation assumes the molecular interaction is due to London dispersion force. Hildebrand termed the mixtures that fit these assumptions as "regular solutions".

    In the paint and coating industry, Burrell[13] and Hansen[14] empirically refined solution theory in an attempt to resolve the inconsistencies due to polar and hydrogen bonding interactions. Hansen[14] empirically resolved the original solubility parameter into dispersive, polar, and hydrogen bonding components. Fowkes[15] attempted to quantify polar and hydrogen components of Hansen's parameter with donor-acceptor interactions, using Drago E and C constants.

    Hildebrand's solubility parameter has been investigated for use in the prediction of flavor sorption by packaging polymers. Haleck et al.[4] and Nielsen[16] used Hildebrand's solubility parameter to explain the sorption behavior of citrus-flavor compounds and polyolefins. Keown[17] had some success using a two-dimensional interaction map based on a modified Hansen approach[18] in the qualitative prediction of flavor sorption; however, this method failed in polymer flavor systems with predominant Lewis acid base interactions. Matsui et al.[19] used the empirical method proposed by Chen[18], a further modification of Hansen's[14] approach that assumes the hydrogen bonding contribution of the δ is negligible. Matsui et al. correlated the solubility coefficient (S) with the distance between the plot of the non-polar (δ_{np}) versus polar (δ_p) component of Hildebrand's solubility parameter, defined as the polymer-flavor compatibility (δ_c). Using semi log plots of d_c vs. solubility coefficients, they obtained good correlations for individual polymers (Figure 1), but correlation was poor over the range of polymers tested.

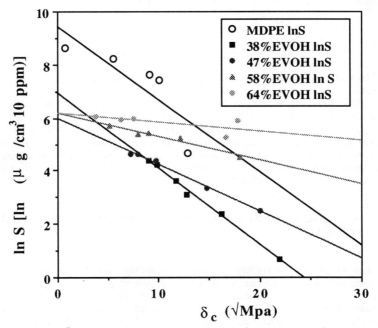

Figure 1 *Correlation of δc and sorption of aroma compounds (Matsui et al.,1992)*

Paik and Tigani[20] attempted to correlate the solubility parameter ($\delta_1 - \delta_2$)^2 with the log of equilibrium sorption. The sorption data obtained using polypropylene (PP), ionomer, and PET polymers showed very poor correlation, which did not improve even when the sorption data for polar compounds were excluded from the data analysis.

3 FLORY-HUGGINS EQUATION

Flory[21] and Huggins[22] independently derived equations, based on the lattice theory of fluids, to describe the free energy change during mixing of a solvent and amorphous polymer. The Flory-Huggins equation[23] (Equation 8), which accounts for both entropic and enthalphic components of mixing in polymer solutions, has been used by many investigators to study the behavior of polymer solutions.

$$\ln a_1 = \ln \phi_1 + [1 - (\frac{1}{m})](1 - \phi_1) + \chi (1 - \phi_1)^2 \tag{8}$$

where subscript **1** represents the solvent
   a = the activity
   ϕ = the volume fraction
   m = the number of segments of the polymer molecule,
and χ = the Flory interaction parameter.

The Flory interaction parameter χ is a function of intermolecular forces. If χ is only related to the heat of mixing and the molecular interaction is dispersive in nature, χ can be estimated from solubility parameters as expressed in equation (9).

$$\chi = \frac{v_1}{RT} (\delta_1 - \delta_2)^2 \tag{9}$$

The volume fraction of flavor sorbed by packaging polymer (ϕ_1) can be calculated from Equation 8.

Paik and Writer[24] compared the ability of the regular solution theory and Flory-Huggins equation to predict flavor sorption by packaging polymers and found (Figure 2) the Flory-Huggins equation to be superior to the regular solution equation in predicting flavor sorption. This suggests that entropic contribution by the molecular size difference is a significant factor in flavor sorption by packaging polymers. While the Flory-Huggins equation can only provide a qualitative prediction of flavor sorption, it is still potentially very useful for selection and design of packaging polymers.

4 UNIQUAC FUNCTIONAL-GROUP ACTIVITY COEFFICIENT (UNIFAC)

Because of difficulty of finding theoretical models that can precisely quantify molecular interactions, semiempirical methods have been applied in the prediction of thermodynamic properties of mixtures. These methods lacked the theoretical rigor of Scatchard-Hildebrand's regular solution theory, Guggenheim's lattice theory, and Flory-Huggin's equation. However, the group contribution methods provided fairly accurate phase equilibrium diagrams for the design of separation processes in chemical production plants. The UNIQUAC functional-group activity coefficient (UNIFAC) model is based on a semiempirical model for liquid mixtures called universal quasi-chemical activity coefficient (UNIQUAC)[25]. This model has a combinatorial contribution to the activity coefficients, essentially due to size and shape differences of the molecules and a residual contribution, and energetic interactions. The combinatorial part is estimated by the Flory-Huggins equation. The Wilson equation (Equation 14 below) that is applied to functional groups is used to estimate the residual contribution part[26]. In a binary mixture, the activity **a** of solvent 1 in polymer 2 is given by:

$$\ln a_1 = \ln a_1{}^C + \ln a_1{}^R \tag{10}$$

The combinatorial part is given by:
$$\ln a_1{}^{C} \approx \ln \phi_1' + \phi_2' + z/2M_1q_1'(1 - \phi_1'/\theta) \tag{11}$$

Two composition variables are the surface fraction ϕ_i' and the segment fraction θ_i';

(12)
$$\phi_i' = \frac{r_i' w_i}{\sum_j r_j' w_j} \qquad ; \qquad \theta_i' = \frac{q_i' w_i}{\sum_j q_j' w_j}$$

where, w_i is the weight fraction of component i.

The pure component parameters are measures of van der Waals volume (q_i) and surface areas (r_i). These two parameters are calculated (Equation 13) from the sum of the molar group volume and group area parameters R_k and Q_k, taken from published data[27].

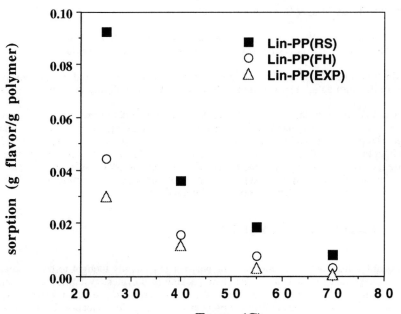

Figure 2. *Comparison of calculated and experimental sorption values of linalool and PP; RS: regular solution theory; FH: Flory -Huggins; EXP: experimental*

$$r_i' = \frac{1}{M_i} \sum_{i\ k} u_k^{(i)} R_k \qquad ; \qquad q_i' = \frac{1}{M_i} \sum_{i\ i} u_i^{(i)} Q_k \qquad (13)$$

where $u_k^{(i)}$ = the number of groups of type k in molecule i
and M_i = the molecular weight of component i.

The residual part is given by:

$$\ln a_1^R = \sum_k u_k^{(i)} [\ln \Gamma_k - \ln \Gamma_k^{(i)}] \qquad (14)$$

where Γ_k = the group residual activity (or activity coefficient)
and $\Gamma_k^{(i)}$ = the group residual activity (activity coefficient) of group k in a reference solution containing only molecules of type i.

$$\ln \Gamma_k = M_k Q_k' [1\text{-}\ln (\Sigma_m Q_m'\Psi_{mk} - \Sigma_m (Q_m'\Psi_{km} / \Sigma_n Q_n'\Psi_{nm})] \tag{15}$$

where Q_m' is the area fraction of the group m, and the sums are over all different groups. Q_k' is the group area parameter per gram such that $Q_k' = Q_k / M_k$. Q_m' is calculated from:

$$Q_m' = Q_m' W_m / \Sigma_n Q_n' W_n \tag{16}$$

where W_m is the weight fraction of group m in the mixture, W_n is the weight fraction of group n in the mixture.

The group interaction parameter Ψ_{mn} is given by:

$$\Psi_{mn} = \exp - (U_{mn} - U_{nn}) / R = \exp - (a_{mn} / T) \tag{17}$$

where W_m =   the weight fraction of group m in the mixture
and W_n =   the weight fraction of group n in the mixture.

a_{mn}, which is called the group-interaction parameter, must be evaluated from experimental phase equilibrium data. Fredenslund[28] shows a wide range of values for parameters a_{mn} and a_{nm}.

Li and Paik[29] used the UNIFAC model to obtain much better predictions of flavor sorption than with the regular solution or the Flory-Huggins equation (Figure 3). Baner and Piringer[30] reported a UNIFAC model estimate of liquid/partition of aroma compounds that proved generally accurate within an order of magnitude. These improved results obtain because the combinatorial part of the UNIFAC equation considers differences in size and shape of molecules and the residual part of the equation relies on experimental rather than theoretical parameters to calculate molecular interactions.

5 CRYSTALLINE FRACTION

The crystalline fraction of a semicrystalline polymer is generally thought to be too compact to accommodate penetrant molecules. The chain packing in polymer crystallites is too dense to sorb even small permanent gas molecules[31]. Although there seems to be an intermediate region, the mass fraction of the intermediate region is almost negligible. Equation 18 can normalize sorption values to amount of flavor only sorbed by the amorphous region of polymers.

$$\omega_1 = \frac{\rho_1 \alpha_a \phi_1}{\rho_a (1 - \phi_1)} \tag{18}$$

Where α_a =   a fraction of an amorphous polymer,
         ρ =   density
and ϕ =   volume fraction

6 CONCLUSION

The regular solution theory and empirical extension of the regular solution theory yielded poor estimates of flavor sorption by packaging polymers, mainly because these theories rely on several invalid assumptions: no excess Gibbs free energy; no free volume change; and dispersive molecular interaction . The main advantage of the regular-solution

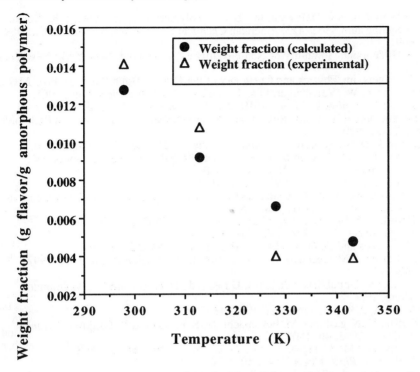

Figure 3. *Comparison of UNIFAC model calculated and experimental sorption values of ethyl alcohol and PP.*

equation is simplicity, which is retained even when the regular solution model is extended to multi-component solutions. The underlying assumptions are more appropriate for small molecular non-electrolytes with predominantly dispersive interactions.

The Flory-Huggins equation estimates flavor sorption better because it takes into account the entropic contribution due to molecular size and shape. However, the Flory-Huggins equation can only provide qualitative predictions of flavor sorption.

The UNIFAC group contribution model quantitatively predicts flavor sorption. This model includes both a combinatorial contribution to the activity coefficients, essentially due to size and shape differences of the molecules, and a residual contribution with energetic interactions based on experimental data. However, the UNIFAC model is limited by proximity effects and a lack of polymer chain conformation. Such limitations must be carefully considered in packaging applications.

References

1. O. Y. Kwapong and J. H. Hotchkiss, *J. Food Sci.* 1987, **52**, 761-763.
2. T. Imai, B. R.Harte and J. R.Giaicin, *J. Food Sci.* 1990, **55**, 158-161.
3. P. Brant, D. Michiels, B. Gregory, K. Laird and R. Day, in: 'Food and Packaging Interactions', J. H. Hotchkiss, Ed., ACS Symposium Series 473. American Chemical Society: Washington, DC, 1991.
4. G. W. Halek and J. P. Luttmann, in: 'Food and Packaging Interactions', J. H. Hotchkiss, Ed., ACS Symposium Series 473. American Chemical Society: Washington, DC, 1991.
5. J. S. Paik and J. A. E. Kail, *Tappi J.* 1992, **75**, 83-87.

6. D. H.Weinkauf and D. R.Paul, In: 'Barrier Polymers and Structure'; J. K. Koros, Ed., ACS Symposium Series 423. American Chemical Society: Washington, D. C., 1989; p 60-91.
7. J. H. Hildebrand, 'Solubility of Non-electrolytes', Rehinhold, New York: New York, 1936.
8. C. E. Rogers, in: 'Physics and Chemistry of the Organic Solid State', D. Fox, M. M. Labes and A. W. Weissberger, Eds., Interscience Publisher: New York, 1965.
9. J. Z. van Laar, *physik. Chem.*, 1910, **72,** 723. Cited in: J. H. Hildebrand , J. M. Prasudnitz and R. L. Scott, 'Regular and Related Solutions',Van Nostrand Reinhold Co: New York, 1970.
10. J. Z. van Laar and R. Lorenz, *Z. anorg. Chem.*, 1925, **146,** 42. Cited in: J. H. Hildebrand , J. M. Prasudnitz and R. L. Scott, 'Regular and Related Solutions', Van Nostrand Reinhold Co: New York, 1970.
11. G. Scatchard, *Chem. Revs.* 1931, **8,** 321-333.
12. J. H. Hildebrand, J. M. Prausnitz, and R. L. Scott, 'Regular and Related Solutions', Van Nostrand Reinhold Co.: New York, 1970.
13. H. Burrell, *Off. Dig. Fed. Paint Varn. Prod. Clubs* 1955, **27,** 726-731.
14. C. M. Hansen, *J. Paint Technol.* 1967, **39,** 104-117.
15. F. M. Fowkes and D. O. Tischler, *J. Polym. Sci.* 1984, **22,** 547-566.
16. T. J. Nielsen, I. M.Jagerstad, R. E. Oste and B. O. Wesslen, *J. Food Sci.* 1992, **57,** 490-492.
17. R W. Keown, Unpublished data. E.I. Du Pont de Nemours and Co., Wilmington, Delaware, 1986.
18. S. A. Chen, *J. Appl. Polym. Sci.* 1971, **15,** 1247.
19. T. Matsui, K. Nagashima, M. Fukamachi, M. Shimoda and Y. Osajima, *J. Agric. Food Chem.* 1992, **40,** 1902.
20. J. S. Paik, and M. A. Tigani, *J. Agr. Food Chem.* 1993, **41,** 806-808.
21. P. J. Flory, *J. Phys. Chem.* 1942, **10,** 51-61.
22. M. L. Huggins, *J. Chem. Soc.* 1942, **64,** 1712-1719.
23. P. J. Flory, 'Principles of Polymer Chemistry ', Cornell University Press: Ithaca, 1953.
24. J. S. Paik and M. S. Writer, *J. Agr. Food. Chem.*, Submitted for publication, 1994.
25. D. S. Abrams, and J. M. Prausnitz, *AIChEJ*, 1975, **21,** 116.
26. E. L. Derr and C. H. Deal, *I. Chem. E. Symp.*, Ser. No. 32. 1969, **3,** 40-51.
27. R. C. Reid, J. M. Prausnitz and B. E. Poling, 'The Properties of Gases and Liquids', 4th ed., McGraw-Hill Book Company: New York, 1987.
28. A. Fredenslund, J. Gmehling, M. L. Michelsen, P. Rasmussen and J. M. Prausnitz, *Ind. Eng. Chem. Process Des. Dev.* 1977, **16,** 450.
29. S. Li, and J. S. Paik, *J. Food Sci.*, Submitted for publication 1994.
30. A. L. Baner and O. G. Piringer, *J. Chem. & Eng. Data*, 1994, **39,** 341.
31. A. S. Michaels, and H. J. Bixler, *J. Polym. Sci.* 1961, **50,** 393.

Supercritical Fluid Extraction (SFE) Coupled to Capillary Gas Chromatography for the Analysis of Aroma Compounds Sorbed by Food Packaging Material

T. J. Nielsen and M. Jägerstad

DEPARTMENT OF APPLIED NUTRITION AND FOOD CHEMISTRY, CHEMICAL CENTER, S-22100 LUND, SWEDEN

1 INTRODUCTION

It is a well established fact that food components can be sorbed by polymeric packaging materials.[1-2] Most such studies have dealt with the sorption of flavour compounds,[3-9] but sorption of other components, such as organic acids and fats, have also been reported.[10-12] This might lead to loss of aroma intensity, or damage to the barrier properties of the package, and a concomitant shelf-life reduction.

There is no standard method available for determining the extent of sorption for a penetrant/polymer system. In previous studies several different methods have been utilized, including gravimetric methods,[13-14] simultaneous distillation-extraction,[15] and conventional liquid extraction procedures followed by a concentration step prior to analysis.[16] These methods are often laborious, time-consuming, and performed using health-hazardous organic solvents. Aroma compounds are present at low concentration levels in foodstuffs, which means that a large amount of sample is needed in order to quantify the sorbed amounts accurately. Furthermore, flavour constituents are usually very volatile and problems with evaporation of the analyte species during the concentration step might occur. In other studies the solubility coefficient has been calculated from values for the permeability and diffusion coefficients, which might result in large errors due to the difficulty in determining the diffusion coefficient. Obviously a rapid and simple method for extracting and analyzing sorbed compounds is needed.

The possibility to use compressed gases as solvents for solids was reported already during the 19th century.[17] It was, however, not until about a decade ago that the use of supercritical fluids for analytical extractions of organic material from complex sample matrices became widely spread. The properties of supercritical fluids are intermediate between those of liquids and gases, and these characteristics make them suitable for extraction purposes.[18] Their low viscosity and high diffusivity make the mass transfer during the extraction rapid. By changing the extraction pressure, the solvent power of a supercritical fluid can be controlled. This important feature is attributed to the fact that the solvent power is related to the density of the fluid, which is altered by a change of state. Another essential factor is that many supercritical fluids are gases at room temperature, which facilitates the concentration of the extract. This characteristic allows the direct coupling of the supercritical fluid extraction with capillary gas chromatography,[19] and thereby less amount of sample is required.

The most commonly used supercritical fluid for extractions is carbon dioxide. It has several advantages over other supercritical fluids. The extreme volatility of CO_2 makes it easy to separate it completely from any solute. Furthermore, it is nontoxic, nonflammable, cheap, and causes no environmental problems. Another important characteristic is the low critical point of carbon dioxide (74 bar, 31°C). This allows extractions to be performed at relatively low temperatures to avoid decomposition of thermally labile analytes.

The aim of this work was to develop a method of supercritical carbon dioxide extraction for measurements of sorbed aroma compounds by plastic packaging material and to collect the extract directly in the gas chromatographic column for subsequent analysis. The method was then used to determine the partition coefficients between selected apple aromas and commonly used food packaging polymers. In another application the method was used to study the sorption of limonene into refillable polyethylene terephthalate bottles.

2 MATERIALS AND METHODS

2.1 Sample Preparation

Ten apple aromas with high aroma values, i.e. concentration/odour threshold, in apple juice,[20] were chosen as sorbates. They were; six esters: ethyl butyrate (etbu), butyl acetate (buac), isopentyl acetate (ipac), ethyl 2-methylbutyrate (e2mb), butyl propanoate (bupr) and hexyl acetate (hxac), two aldehydes: hexanal (hxal) and trans-2-hexenal (t2hx), and two alcohols: isopentanol (ipol) and hexanol (hxol). An aqueous solution of 10 ppm of each of the selected volatiles was prepared. The sorption of the aroma compounds by five different polymers commonly used for food packaging was measured by the supercritical fluid extraction/gas chromatography (SFE/GC) technique. The polymers studied were: low density polyethylene (LDPE), linear low density polyethylene (LLDPE), polypropylene (PP), nylon 6 (PA) and polyethylene terephtalate (PET). Separate strips of polymer films were stored in 5 ml of the solution in glass ampoules, sealed by flame. The ampoules were stored in the dark for one week prior to analysis.

In the next experiment, a carbonated orange flavoured soft drink was stored in refillable PET bottles for 12 weeks at 4°C and 25°C. The sorbed amount of limonene, an orange peel oil, was analyzed after 1, 2, 4 and 12 weeks of storage. After 4 weeks the bottles were washed with sodium hydroxide solution of 3.0 % (w/v), containing 0.25 % (v/v) detergent, at 60°C, and the remaining amount of limonene in the polymer was determined.

2.2 Equipment

The SFE/GC apparatus was arranged according to Figure 1. It was essential to have a carbon dioxide of high purity (99.998 %) to minimize its interference with the chromatographic analysis. The cooling bath of -15°C was necessary to liquefy the carbon dioxide since the pump used could pump only liquids. The gas chromatograph was a Varian 3400 with an additional cooling device for the column oven. The SFE outlet restrictor had an internal diameter of 25 μm. All connections were of stainless steel with internal diameter of $^1/_{32}$ inches.

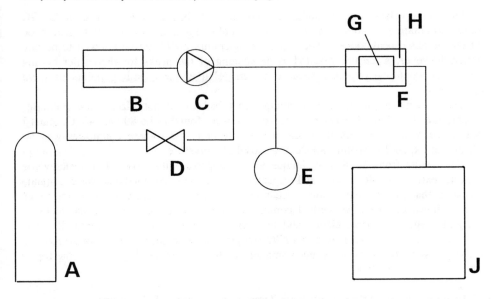

Figure 1 *Equipment for SFE/GC. (A) carbon dioxide flask, (B) cooling bath, (C) pump, (D) pressure regulator, (E) pressure meter, (F) water bath, (G) extraction cell, (H) immersion heater, (J) gas chromatograph*

2.3 Analysis

A weighted polymer sample was wiped dry and placed in the extraction chamber, which was immersed into a water bath where the temperature was controlled by an immersion heater. The capillary restrictor was then inserted into the GC column through the on-column injection port. During the extraction the GC column oven had to be cooled in order to cryofocus the extracted compounds, but let the CO_2 pass. Carbon dioxide was pumped through the extraction cell, and the extraction was assumed to be initiated from the moment the pressure reached the desired value. After the extraction the restrictor was taken out and the CO_2 left in the column was let out. Subsequently, the column was flushed for two minutes with the carrier gas before the column was rapidly heated to 40°C and the chromatographic analysis was performed with the conditions as follows: Column Supelcowax 10 (60 m x 0.25 mm), He carrier at 1.0 mL/min, H_2 at 30 mL/min, air at 300 mL/min, He make-up at 30 mL/min, injector at 40°C with increase of 160°C/min to 200°C, flame ionization detection at 260°C and column oven from 40°C to 150°C at 5°C/min.

3 RESULTS AND DISCUSSION

3.1 Optimization of Supercritical Fluid Extraction [21]

Several factors were found to be of importance for yielding a complete extraction of the analytes. Both the optimal hardware design and the optimal SFE conditions had to be established.

Due to the volatility of the studied compounds, they had to be cryofocused in the GC column to obtain a satisfying chromatogram. A column temperature of -50°C gave good chromatographic peak shapes. The column temperature could not be allowed to be any higher, however, since this resulted in broadening of the peaks for the earlier eluting species as would be expected because of the relatively low boiling points of these substances.

The parameters of the SFE technique that had to be optimized were pressure, temperature, and time. The optimal adjustments were found to be 80 bar, 40°C, and 15 min, respectively. The gas flow out of the restrictor under these conditions was 50 mL/min of expanded carbon dioxide at atmospheric pressure.

The results from the SFE technique were compared with those from a methylene chloride extraction. Recovery and repeatability of the two methods were similar, indicating that the two extraction procedures were just as effective. Considering other aspects, however, the SFE method compared favourably to the conventional method, especially regarding time, effort, and amount of sample needed to perform the two separate extractions. A complete SFE/GC analysis could be accomplished in 40 min, in one single step and it needed a much smaller amount of sample (1-2 %) than the liquid solvent extraction method.

3.2 Influence of Sorbate and Sorbant Type on the Extent of Sorption [22]

The amount of a compound sorbed by a polymer can be expressed by calculating the partition coefficient, K, as the concentration of the substance in the polymer phase divided by the concentration of the substance in the aqueous phase. The partition coefficients between the analytes and the separate polymers in this study are summarized in Table 1.

There were major differences between amounts of aromas sorbed by the different polymers. Generally, the polyolefins, i.e. LDPE, LLDPE and PP, sorbed larger quantities of the analytes than did the polar polymers, PA and PET. Amounts sorbed of the different compounds also varied a great deal, the esters and aldehydes being sorbed more than the alcohols in all polymer films, apart from PA. Further, it was observed that the molecular size had a great impact on the K values, larger molecules being sorbed to a larger extent.

In an effort to explain the obtained results the solubility parameter, δ, has to be considered.[23] A comparison of the δ values of a polymer and an organic substance gives an indication of the solubility. The smaller the difference between the δ values the greater the solubility. Other factors important in prediction of solubility include polarity and hydrogen-bonding character of the sorbate and the sorbant.

The esters and aldehydes had δ values very near to LDPE, LLDPE and PP, which partly explains their large solubility in these polymers. The differences between the esters might be attributed to the length of the carbon chain - the longer the chain, the less polar and the easier the compounds were sorbed by the non-polar polyolefins. The alcohols had much larger δ values than the polyolefins, which might explain why they were less sorbed than the other analytes. PET, on the other hand, had a solubility parameter near to the δ values of the alcohols. The reason why they were not sorbed to a greater extent by this polymer might be their strong hydrogen-bonding character, which PET did not exhibit. The difference between the δ values of PET and of the esters and aldehydes was too large for any significant sorption to occur. The δ value of PA was much higher than the solubility parameters of the esters and aldehydes but near the δ values of the alcohols, especially hxol. Further, both PA and the alcohols had a strong hydrogen-bonding character, which explains why PA showed the highest affinity for hxol.

Table 1 *Partition Coefficients of Selected Apple Aromas Between Polymer Films and Aqueous Solution at 25°C after 1 Week of Storage*

Polymers	Aroma compounds									
	etbu	buac	ipac	e2mb	bupr	hxac	hxal	t2hx	ipol	hxol
LDPE	2.3	2.5	5.3	6.4	7.0	24.8	3.7	11.6	0.6	0.4
LLDPE	1.6	1.7	3.6	5.2	4.2	19.0	1.5	11.8	0.4	0.3
PP	4.4	5.5	9.3	11.3	18.8	45.1	6.4	26.2	1.1	0.6
PA	0.5	0.6	0.6	0.5	1.1	2.0	1.6	1.0	0.3	0.8
PET	0.6	0.7	0.5	0.3	1.3	3.8	1.1	0.9	0.3	0.3

Means of two separate studies with five replications each. Relative standard deviation less than 12 %. The aroma substances and polymers are as listed in the text, section 2.1.

3.3 Limonene Sorption by Refillable PET Bottles [24]

The levels of limonene sorbed into the PET wall during 12 weeks storage at the two test temperatures are displayed in Table 2. The sorption continued during the entire experiment, which could be expected because of the low diffusion coefficient for limonene in PET, and the thickness ($\sim$ 1 mm) of the polymer. There was a significant difference between limonene amounts sorbed by bottles stored at 4°C, and those stored at 25°C, with the sorption much greater at the higher temperature. This may indicate a different equilibrium constant at the two temperatures, a slower diffusion process, or a combination of both. The relative differences between the sorption levels at the two temperatures decreased with storage time. After 1 week PET bottles stored at 25°C had sorbed seven times more limonene than the bottles stored at 4°C, but only three times more after 12 weeks. This indicates a slower diffusion process at 4°C and the difference might be less after an even longer storage time. After 12 weeks storage at 25°C the PET bottles had sorbed 9.9 μg limonene/g plastic. Assuming the sorption was uniform in the entire bottle a total of 0.41 mg limonene was sorbed by the polymer during this period. This corresponds to 1.4 % of the total limonene content present in the soft drink.

The washing experiment produced some interesting results. Only 22 % of the limonene sorbed into the PET bottles was removed by washing at 60°C with the sodium hydroxide solution. The fact that considerable amounts of aromas were still left in the polymer after washing was also confirmed by a strong smell of orange from the bottles. This is, of course, unacceptable if the bottles are to be refilled with any other product. However, in commercial washeries the washing is probably much more efficient. In this study, the washing solution was simply poured into the bottles and held at 60°C. The bottles were shaken a few times during the cleaning, but mechanical effects were not as strong as in commercial washing where the solution is flushed into the bottles. In this investigation, it was not possible to wash the bottles to simulate better an industrial process. Even if washing was not comparable, results indicated that sorbed aroma compounds are not easily removed from the bottles once they have been sorbed. It is highly probable that sorbed substances might be retained in bottles even after a more severe washing.

Table 2 *Sorption of Limonene ($\mu g/g$ plastic) from Orange Flavoured Soda by PET during 12 Weeks of Storage at 4°C and at 25°C*

Temperature	Week 1	Week 2	Week 4	Week 12
4°C	0.65	0.98	1.35	3.41
25°C	4.50	4.95	6.62	9.90

Means of triplicate analyses. Relative standard deviation less than 10 %

References

1. P. Durr, U. Schobinger and R. Waldvogel, *Lebensm.-Verpackung*, 1981, **20**, 91.
2. M. R. Marshall, J. P. Adams and J. W. Williams, "Aseptipak", Schotland Business Research Inc., Princeton, NJ, USA, 1985, p 299.
3. Z. N. Charara, J. W. Williams, R. H. Schmidt and M. R. Marshall, *J. Food Sci.*, 1992, **57**, 963.
4. A. P. Hansen and D. K. Arora, "Barrier Polymers and Structures", American Chemical Society, Washington, DC, USA, 1990, Chapter 17, p. 318.
5. T. Ikegami, K. Nagashima, M. Shimoda, Y. Tanaka and Y. Osajima, *J. Food Sci.*, 1991, **56**, 500.
6. J. B. Konczal, B. R. Harte, P. Hoojjat and J. R. Giacin, *J. Food Sci.*, 1992, **57**, 967.
7. S. M. Mohney, R. J. Hernandez, J. R. Giacin, B. R. Harte and J. Miltz, *J. Food Sci.*, 1988, **53**, 253.
8. M. G. Moshonas and P. E. Shaw, *J. Food Sci.*, 1989, **54**, 82.
9. G. Sadler and R. Braddock, *J. Food Sci.*, 1991, **56**, 35.
10. W. D. Bieber, K. Figge and J. Koch, *Food Add. Contamin.*, 1985, **2**, 113.
11. K. Figge, *Progr. Polym. Sci.*, 1980, **6**, 187.
12. G. Olafsson, M. Jägerstad, R. Öste, B. Wesslen and T. Hjertberg, *Food Chem.*, 1993, **47**, 227.
13. G. W. Halek and M. A. Meyers, *Packaging Technol. Sci.*, 1989, **2**, 141.
14. C. H. Mannheim, J. Miltz and A. Letzter, *J. Food Sci.*, 1987, **52**, 737.
15. M. Shimoda, T. Ikegami and Y. Osajima, *J. Food Sci. Agric.*, 1988, **42**, 157.
16. O. Y. Kwapong and J. H. Hotchkiss, *J. Food Sci.*, 1987, **52**, 761.
17. J. B. Hannay and J. Hogarth, *Proc. Roy. Soc.*, 1879, **29**, 324.
18. F. Kelnhofer and G. Tittel, *Int. Chrom. Lab.*, 1993, **13**, 16.
19. S. B. Hawthorne, M. S. Krieger and J. D. Miller, *Anal. Chem.*, 1988, **60**, 472.
20. L. Poll, *Lebensm. Wiss. Technol.*, 1988, **21**, 87.
21. T. J. Nielsen, I. M. Jägerstad, R. E. Öste and B. T. G. Sivik, *J. Agric. Food Chem.*, 1991, **39**, 1234.
22. T. J. Nielsen, I. M. Jägerstad, R. E. Öste and B. O. Wesslen, 1992, *J. Food Sci.*, **57**, 490.
23. J. H. Hildebrand and R. L. Scott, "The Solubility of Non-Electrolytes", Reinhold, New York, NY, USA, 1959.
24. T. J. Nielsen, *J. Food Sci.*, 1994, **59**, 227.

The Influence of Food and External Parameters on the Sorption of Aroma Compounds in Food Contact Polymers

A. Leufvén, F. Johansson, and C. Hermansson

SIK, THE SWEDISH INSTITUTE FOR FOOD RESEARCH, BOX 5401, 40229 GÖTEBORG, SWEDEN

1. INTRODUCTION

Every food packaging material possesses certain barrier characteristics against compounds that are potentially harmful to the quality of the packed food. These compounds could be permanent gases, such as oxygen, and water vapour, or they could be organic compounds that damage or enhance the aroma of the food.

The barrier properties of a specific packaging material can be tested. These tests are, however, almost always performed with the packaging material in its virgin condition, disregarding the fact that something will be packed in contact with the material. Sometimes the effects of external parameters, such as that of humidity on oxygen transmission through the material, have been taken into account, but the effect of food parameters on the packaging material is unfortunately very seldom considered in the design of a packaging solution.

This presentation will focus on sorption phenomena. Sorption may affect the food quality *per se* by withdrawing aroma compounds from the food. This can be illustrated by aroma components from coffee sorbed into the packaging material. Flavour carry over in refillable PET bottles is another example that will be covered in the next presentation [1]. Together with diffusion, sorption determines permeation. Sorption can also change the properties of the polymer. This will have an effect on the permeation, diffusion and additional sorption of compounds into the polymer giving the polymer new barrier characteristics.

2. METHODS

All permeation measurements on aroma compounds (i.e. aldehydes, alcohols, *etc.*) were made using a specially designed permeation cell in which polymer films could be positioned [2]. The aroma compounds were presented as vapours, collected on adsorbent material after passing through the film, and analysed chromatographically after thermal desorption from the adsorbent.

A modified [3] version of the supercritical carbon dioxide method developed by Nielsen *et al.* [4] was used in all sorption studies to remove aroma compounds from the polymer.

The determination of oxygen transmission through polymer films at different humidities was made using a Mocon Oxtran apparatus.

3. RESULTS

3.1. Temperature

The influence of temperature on the permeation, sorption, and diffusion of aldehydes was studied [2] in different "food-wrap" films, both laminates and those made from a single polymer. In general the solubility of aldehydes in the polymers was influenced by temperature in a way similar to the solubility of gases in liquids with decreasing solubility at increasing temperatures. The diffusion, on the other hand, increased with increasing temperature. Solubility (sorption) proved to be more important than diffusion in these cases, resulting in a greater aldehyde permeation through the polymer film at lower temperatures.

3.2. Humidity

The humidity also has an influence on the interaction between aroma vapours and polymers. The influence is different for different aroma vapours and different polymers. This is exemplified by the solubility and permeability of aldehydes and alcohols [5].
In LLDPE, the solubility of both the alcohol and the aldehyde was at its lowest at an intermediate humidity. The humidity-induced changes in permeability were different for the alcohol and the aldehyde. The alcohol permeability in the LLDPE decreased with increased humidity but the aldehyde permeability was at its maximum at an intermediate humidity.
In the more polar EVOH polymer the interactions between the polymer, water, and aldehyde or alcohol resulted in different solubility and permeability characteristics. The solubility of the polar alcohol increased drastically at the higher humidity. However, the permeability of the alcohol through the EVOH polymer was only minimal at this high humidity. The opposite was true for the aldehyde where the change in solubility was slight but the permeability increased with increased humidity. The results from these experiments are further discussed in the poster session [6]

3.3. pH

The pH of the food affects the sorption of aroma compounds into the packaging material. We have studied this at SIK [7], using aroma compounds (i.e. *trans*-2-hexenal, 2-heptanone, 6-methyl-5-hepten-2-one, 6-methyl-5-hepten-2-ol, and limonene) from tomato juice as a model system. The interactions that determine the amount sorbed into the polymer are complex. This is probably due to time dependent changes in both the polymer and the aroma compounds induced by the pH. Hexenal will be used as an example of pH-influenced sorption of aroma compounds into polymers. In PET, the sorption of hexenal increased with increased pH, but in both PE and EVOH the sorption of hexenal into the polymers was greater at pH 5 than at pH 3 and pH 7. These experiments were performed at room temperature using a water-ethanol mixture with components from tomato juice added. On adding the aroma compounds to authentic tomato juice instead, the interactions proved to be even more complex due to the introduction of an additional number of possible equilibria into the system. In the hexenal example, the amount sorbed into the PET polymer was smaller at pH 7 if tomato juice was present than if the pure model system was used. The sorption of hexenal into PE and EVOH polymers was less affected by tomato juice than the sorption into PET.
The complexity of the system was further demonstrated by the fact that a longer contact time between the polymer and the solution containing the aroma compounds did not always result in greater sorption of aroma compounds into the polymer at all pH values. This can be exemplified by the amounts of different aroma compounds extracted from PET stored at pH 3 from one day to two weeks (Figure 1). At this pH, the amounts that could be extracted from the polymer generally increased with increased storage time. Limonene, however, was present in largest amounts already after one day of storage and then the amounts decreased with increased storage time. This phenomenon could not be detected after storage at pH 5 where the amount of limonene extracted from the polymer increased with increased storage time.

3.4. Sorption of Other Compounds

3.4.1. Aroma Compounds

Sorption of additional aroma compounds into a polymer may affect the sorption of other aroma compounds into the polymer. We have exemplified this using hexanal and pentanal vapours [8]. Compared with pentanal almost twice the amount of hexanal was extracted from polyethylene after exposure to similar amounts of the two aldehydes separately. However, when the compounds were presented together, the amount of pentanal sorbed into the polymer was increased but the amount of hexanal remained almost constant.

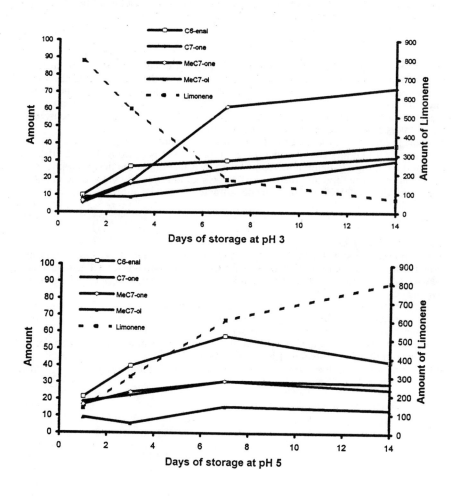

Figure 1. *The amount, in μg cm³, of several aroma compounds from tomato juice sorbed in polyethylene after storage in water-ethanol mixtures with different pH.*

3.4.2. Vegetable Oil

A component of the food, such as vegetable oil, can be sorbed into the packaging material. This sorption can change the barrier characteristics of the polymer. Our experiments [9], which are further discussed in a later chapter [6], have shown that the sorption of oil can reduce the efficiency of the polymer as an oxygen barrier. This was the case for polypropylene for instance. The relative humidity also affects the efficiency of some polymers as oxygen barriers. As our experiments have shown, the sorption of oil into polymers may have different effects on the transmission of oxygen through the polymer at different humidities. This was the case for APET. At dry conditions, both the virgin and the oil-containing polymer were equally efficient as oxygen barriers. At low to intermediate humidities, the oil-containing polymer was a better oxygen barrier than the polymer containing no oil. However, at intermediate to high relative humidities the virgin polymer was the better oxygen barrier.

4. CONCLUSIONS

The factors that influence the ability of food packaging materials to preserve the quality of the food are numerous, complex, and interacting. Swelling of the polymer, changes in affinity between polymer and permeant/sorbant, and "loss of space" in the polymer are probably important factors governing these interactions. A lot of experimentation, thinking, and verification of hypotheses has to be carried out before a comprehensive theory that takes all the different ways of interaction between polymer, food, environment, permeants, sorbants, *etc* into account can be put forward.

However, even without an extensive theoretical framework it is obvious that the effect of the food on the packaging material has to be taken into account in the design of a food packaging solution. The food and the food package should always be viewed as an entity that determines the quality of the food during storage.

References

1. U. Stöllman, 1995, *This volume*
2. A. Leufvén and U. Stöllman, 1992, *Z. Lebensm. Unters. Forsch.*, **194**, 355.
3. F. Johansson, A. Leufvén and M. Eskilson, 1993, *J. Sci. Food Agric.*, **61**, 241.
4. T.J. Nielsen, I.M. Jägerstad, R.E. Öste and B.T.G. Sivik, 1991, *J. Agric. Food Chem.*, **39**, 1234.
5. F. Johansson and A. Leufvén, 1994, *J. Food Sci.*, In press.
6. F. Johansson and A. Leufvén, 1995, *This volume*.
7. A. Leufvén and C. Hermansson, 1994, *J. Sci. Food Agric.*, **64**, 101.
8. A. Leufvén, F. Johansson and C. Hermansson, 1994, *In* H. Marse and D.G. van der Heij 'Trends in Flavour Research', Elsevier Science, 431.
9. F. Johansson and A. Leufvén, 1995, *Pack. Technol. Sci.*, Submitted.

Aroma Transfer in Refillable Pet Bottles – Sensory and Instrumental Comparison

U. Stöllman

SIK – THE SWEDISH INSTITUTE FOR FOOD RESEARCH, P.O. BOX 5401,
S-402 29 GÖTEBORG, SWEDEN

1 BACKGROUND

As a result of intense political discussions about the use of packages and their environmental impact, the Swedish Brewer's Association introduced in June 1991 the first refillable polymer packaging system on the Swedish market. It was a standard 1.5 litre PET bottle for soft drinks. Today we still have this standard bottle and all soft drinks would be filled in the same type of refillable PET bottle, with the exception of Coca Cola and Pepsi Cola which have their own profile bottle. Similar systems are on the market in countries such as Norway, Finland, Denmark and Holland. Some of them have systems with more than one bottle size.

In 1993, when the exemption for disposable bottles ended, there were new discussions about the refillable system. The aspects then concerned competition restrictions. The washing systems and the long-distance distribution of the bottles to the plants were too expensive for the smaller soft drink fillers. This meant that only the biggest breweries could afford these types of investment and as a consequence of this they could get advantages to dominate the soft drink market.

The final political decision resulted in there now being two systems on the market:
1. The refillable system
2. Disposable PET bottles for material recycling.

The trend today, however, is that also the smaller fillers are interested in joining the refillable system due to the reluctance of the retailers to handle the enourmous amount of disposable bottles.

2 PRODUCT SAFETY ASPECTS

All refillable systems involve certain problems in terms of product safety, and it is essential that both hygienic and legislative aspects are taken into account.

The washing procedure is a particularly critical step in the refilling of PET bottles, due to the limited heat resistance of the packaging material. To avoid shrinkage of the bottle the temperature must not exceed 58-59°C. This requires longer contact times between the lye and the bottle during the washing procedure.

To eliminate bottles which have been used to store products other than soft drinks, for instance petroleum products, cleaning materials, specific detectors, so-called sniffers, are installed on the line. The sniffer, which is sensitive to foreign odours, will reject contaminated bottles before the washing process. The development of the sensitivity and selectivity of the sniffer has been very rapid in recent   years and today several different systems are in use. When it comes to the sensory quality of the packed products it is essential to ensure that aroma compounds from a product in the bottle are not absorbed by the PET material and transferred to the next product filled in the bottle.

3 OFF-TASTE ANALYSES

Before the different types of soft drink from the breweries and other fillers are accepted for filling in refillable bottles they must be tested and approved. This applies to all new products and also any formula changes to previously tested prdocuts.

The best way of evaluating the risk of aroma transfer when a bottle is refilled with different drink is to use sensory analysis. As an independent institute with considerable experience of sensory and chemical analysis, SIK has been commissioned to perform all the SENSORY ANALYSES of soft drinks packed in refillable bottles, in order to assess the risk of taste transfer between products when a returnable PET bottle is refilled with a different soft drink.

3.1 **Washing, Filling and Storage**

Test packaging, washing and storage are carried out according to well-defined instructions provided by the Swedish Brewers' Association.

The test procedure for a particular soft drink is carried out in the following way, see Figure 1.

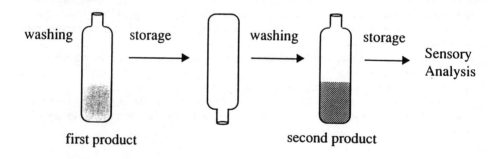

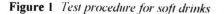

Figure 1 *Test procedure for soft drinks*

After washing of the new bottle it is filled with *the first product, the test product,* that could be any soft drink that is to be tested and approved for the system. Bottles that will be used as reference samples are here filled with mineral water. The bottles are then stored for 3 weeks at 30°C in darkness. The bottles are then emptied, washed again and filled with the *second product* on which the assessment of possible aroma transfer will be made. The products filled the second time are soft drinks of a citrus type, cola type and mineral water, respectively. On the basis of earlier experiences we chose to work with a limited number of products as it is impossible to test all combinations. The bottles filled with the second product are stored at 30°C for another 4 weeks before sensory analyses are performed

3.2 SensoryTest Design

When working with sensory analysis, as well as all other kinds of analysis, the choice of method is crucial, as is the panel used in order to get reliable results. When working with aroma transfer problems you have to work with off-tastes with very low intensities.

If you are interested in determining whether there are any differences between two samples, the most commonly used method is the *triangular test*. Three coded samples are used, two of which are identical and one is different from the others. The panellist has to pick out the odd sample and even if no difference is noted, one sample has to be marked as the "odd" one. This means that there is a 33% chance of giving the right answer. This method is used by many companies and institutes today when this type of aroma transfer is tested. However, caution is required as there are some drawbacks. The only question to be answered is *whether there is a difference or not* between two samples but nothing about the type of difference is noticed. As a consequence of this, the method could indicate significant differences due to other properties than the actual off-flavour differences, such as colour differences, differences in carbonation or other differences that may be noticed and which have nothing to do with the aroma transfer. To provide reliable results this method also requires *many judges,* about 20 persons. *The size of the difference can not be established either.*

Instead of this method we use a modified intensity test. Two samples are assessed, one reference and one coded sample. The reference sample is taken from a bottle in which the first product was mineral water, which means that the reference is completely free from off-taste (by definition). The coded sample and the known reference sample are served in pairs and the judge first tastes the reference sample, then the coded soft drink sample and marks the total intensity of the off-taste in the sample (compared to the reference which is quite neutral). At each session, different products are assessed at the same time but always in relation to a reference sample.

To ascertain that the judges are able to discriminate between the sample and the reference, certain coded samples are references without off-taste, so-called blind samples. The panel consists of eight to ten specially selected and trained judges. To retain the assessment ability of the panels during the test period the judges are trained continuously according to specially designed training schemes.

3.3 Evaluation and Results

Based on our knowledge from empirical studies of off-taste in soft drinks, five classification levels (1-5) were established on the basis of the panellists' judgements of the off-taste in the soft drinks. (5 = all judges are aware of a distinct off-taste; 1 = none of the judges is aware of an off-taste). These findings were used as a tool in the assessment of the objective mathematical evaluation method. The objective evaluation is based on mathematical statistical hypothesis tests, so-called similarity tests, which means similar to the reference. The "approved" or "failed" limits depend on how strictly the requirements in the statistical evaluation model are set (i.e. where you want to set your limits for your products).

The intensity of the off-taste in the sample is obtained by comparing with the reference. The "noise" in the assessments is calculated using the blind samples.

As mentioned before, the test product is judged for the transfer of aroma to soft drinks such as citrus-based and cola-based products. For each of these, the intensity of the aroma transfer is evaluated separately. The test product will be approved for filling in returnable PET bottles only if it is approved in all the tests. Their results show that the test products could give different reactions when tested in cola- or citrus-based product, see Figure 2.

Products 3 and 8 transfer hardly any off-taste at all to any of the second products, while the opposite is noticed for products 2 and 10. Products 1 and 2 gave more off-taste in the citrus product while the opposite was seen for products 9 and 10, which gave a stronger off-taste in the cola.

The reasons for this could be both chemical and sensory. Some of the products may be more similar in composition to the cola or the citrus products.

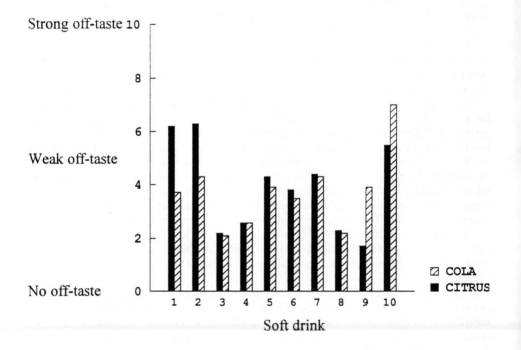

Figure 2 *Aroma transfer from different products to cola- and citrus-based soft drinks*

Our studies show that:
o   Certain products are more likely to transfer aroma than others. Products with very intense and strong aroma, such as certain citrus and cider fruit drinks, may give rise to problems in returnable PET bottles.

This does not mean that no such products could be filled, and both ciders and citrus drinks are used in the system.

o Certain products are more susceptible to aroma uptake than others. Highly taste-sensitive products, such as mineral water, can only be filled in new bottles or in bottles which have previously contained water.

3.4 Chemical Investigation

Sensory evaluation of aroma transfer must, of course, be supplemented by chemical studies in order to determine which aroma compounds are sorbed by the PET material, enabling them to migrate to the next product in the bottle.

If this could be done, the next step would be to investigate the role of these compounds as aroma components and, if possible, replace or reduce them or find methods to eliminate them from the bottle.

For the analysis of chemical components sorbed in packaging materials we have applied two different techniques; headspace analysis (gas stripping) of the inside of the bottle at elevated temperature and extraction of the material by supercritical carbon dioxide. These techniques give different quantitative compositions of aroma compounds desorbed from the wall, depending on the properties of the components.

The amount of material that can be analysed also differs greatly between the two methods. To illustrate the different techniques we investigated what type of flavour compounds were identified as having been sorbed and to migrate from bottles previously filled with an apple cider product. The bottles were analysed after repeated rinsing with warm water.

The dominating compounds identified to be migrating from the bottle wall by using the gas stripping method were esters, some aldehydes and alcohols. Most of them are characterised as the dominating flavour compounds in the cider drink itself. Limonene was only found to be a minor component.

When supercritical CO_2 extraction of pieces of the wall was used, the quantitative pattern, was different. Limonene was found to be the most effective aroma compound extracted from the wall. It dominated and was present in much higher concentration than the other aroma compounds, such as esters, aldehydes and alcohols. Limonene is often used as a model compound in sorption studies and is effectively extracted using supercritical CO_2.

Both sorption and desorption processes are very complex and specific and have to be properly understood when working with prediction of aroma transfer. The possibilities and limitations of the analytical methods must be evaluated and compared. Sensory and instrumental methods used together provide a suitable tool for evaluating compounds which could cause aroma transfer problems. Other parameters, that also play a role for the sorption, are the pH value of the packed product and the structure (cristallinity) of the material. Information of this kind represents an important tool in the development of new soft drink aromas, bottles and washing procedures.

4 CONCLUSION

Our investigation has shown that flavour transfer problems can occur when different soft drinks are refilled in PET-bottles. When new refillable systems for food products are discussed the sensory aspects as well as hygienic and toxicological aspects must be taken into account. A great deal of work remains to be done, both on the choice of material and the washing procedure, before safe refill systems which do not require testing for aroma transfer effects will be available.

Taste Recognition Threshold Concentrations of Styrene in Foods and Food Models

J.P.H. Linssen, J.L.G.M. Janssens, J.C.E. Reitsma, and J.P. Roozen

WAGENINGEN AGRICULTURAL UNIVERSITY, DEPARTMENT OF FOOD SCIENCE, PO BOX 8129, 6700 EV WAGENINGEN, THE NETHERLANDS

1 INTRODUCTION

Packaging materials are potential sources of off-flavour in packed food products. Polystyrene is frequently used as material for the packaging of foods. Typical examples are yoghurt and dessert packaging, foamed trays for meat and crystal clear trays for the packaging of salads and vegetables. Polystyrene contains residual monomer in detectable amounts. Styrene monomer can only partly be removed from the polymer by extrusion of the packaging material. Styrene is able to migrate into food products and may affect product quality because of its unpleasant plastic-like chemical odour and/or taste. Several authors reported the presence of styrene in food products packed in polystyrene packaging material, e.g. 180 ppb in chopped peel,[1] up to 245 ppb in sour cream[2] and even up to 3.59 ppm in mousse.[3] An off-flavour in chocolate and lemon cream cookies[4] packed in polystyrene trays and milk and plain chocolate in contact with polystyrene[5] was detected by different taste panels. The intensity of off-flavour depended on the level of residual monomer in the polystyrene packaging material, type of food matrix and contact time.

Threshold values are important parameters for (off)-flavour perception. (Off)-flavour perception is determined by the nature and quantity of the flavour compound and the availability of such compound to the sensory system as a function of time. Also the food matrix plays an important role in the perception of flavours. As shown in Table 1, Jenne[6] reported that the recognition threshold concentrations of styrene increased with increasing fat content of the food products.

Table 1 *Recognition Threshold Values of Styrene in Food Products*

Food Product	Recogniton Threshold Value (ppm)
Tea	0.2
Nectar	0.2
Low fat milk	0.3
Yoghurt	0.5
Full fat milk	1.2
Vanilla custard	1.5
Cream	6.0

2 MATERIALS AND METHODS

2.1 Samples and Sample Preparation

Water, oil in water (O/W) emulsions, yoghurts and yoghurt drinks were used as medium in the experimental part of the study. O/W-emulsions were prepared by dispersing 3, 10, 15, 20, 25 and 30% corn oil in water using 1% sodium stearoyl-2-lactylate (Admul SSL 2004) as emulsifier. Polyethylene packed yoghurts, containing 0.1-3% fat, and sugared (ca 10%) and flavoured (natural, strawberry, peach) yoghurt drinks, containing 0.1% fat, were bought in a local store in Wageningen (The Netherlands). Test series were prepared by spiking the samples with different amounts of styrene.

2.2 Sensory Evaluation

Taste recognition threshold concentrations (TRTC) of styrene were determined by using two different panels of respectively 48 assessors for tasting water and the O/W-emulsions and 24 assessors for tasting the yoghurts and the yoghurt drinks. The assessors were all students of the Wageningen Agricultural University and aged 20-25 years old. TRTC of styrene was defined as its concentration, which was recognized in a sensory test by 50% of the assessors. In the sessions for tasting, the emulsions standard samples of emulsions with and without styrene were presented first to the assessors for assuring recognition of the taste quality of styrene. In the sessions for tasting the other samples, water containing 300 ppb styrene was used as standard. Each test series consisted of 9 solutions, which were offered in random order to the assessors as 15 ml samples in 100 ml glass bottles covered with aluminium foil and tightly closed with screw caps. The assessors were asked to keep a sample 20 seconds in the mouth or shorter until styrene was recognized. The time interval between tasting of samples was 1 minute. During that time, the assessors were asked to rinse their mouths with Seven-Up and water and to eat crackers for recovering their taste abilities.

Data were collected by using a computer interactive interviewing system (Ci2 system, Sawtooth Software, Inc., Ketchum, ID, USA) on field disks on portable computers, which were placed in sensory evaluation booths. Data from the field disks were accumulated onto a hard disk (HP Vectra/HS) and converted into a SPSS-PC[+] data file for further elaboration.[7]

2.3 Styrene Analysis

In the yoghurt and yoghurt drinks samples the exact concentration of styrene was determined just before tasting by azeotropic distillation with methanol, followed by hexane. The hexane extracts were analysed with a gas chromatography (Carlo Erba Model 4160), equipped with a flame ionisation detector and a cold on-column injector, on a fused silica capillary column (30 m x 0.32 mm i.d.; DB 1701, J&W Scientific, Folsom, CA, USA). Calibration curves were prepared by adding various amounts of styrene to the samples.[8]

3 RESULTS AND DISCUSSION

The concentrations of styrene in food and food models used in this study for determining the TRTC of styrene are presented in Table 2. The 50% TRTC's for each sample can be calculated from the linear regression equations obtained by plotting the proportions of positive responses of the panel in the recognition test, possibly converted into Z-values using a conversion table, versus the logarithm of the stimulus concentration, i.e styrene. The calculated 50% TRTC's from these regression equations are presented in Table 3.

Water is a very sensitive product for acquiring a taint and the lowest TRTC of styrene was found. Table 3 shows that the TRTC's for the emulsions and yoghurts increase with increasing fat content of the products and so making styrene less noticable in high fat products. The relationship between the amount of an O/W-emulsion and the TRTC value is demonstrated in Figure 1. The TRTC increases linearly with increasing fat content in the O/W-emulsions. The linear regression equation calculated is as follows: 50% TRTC (ppm) = 0.0068 x oil concentration (%) + 0.035. Linssen et al.[7] show that also the TRTC values for water and yoghurts are in good position in respect to the regression line of Figure 1.

The artificial flavoured yoghurt drinks contain about 10% sugar. The TRTC value for the yoghurt drink naturel was found to be twice as high as the value found for the low fat yoghurt. The presence of sugar masks the TRTC of styrene. The presence of flavourings, however, influences the TRTC values of styrene in yoghurt drinks only very slightly. The unflavoured yoghurt drink natural shows only a slightly decreased TRTC value compared with the strawberry and peach flavoured yoghurt drinks (Table 3).

Table 2 *Concentrations (ppm) of Styrene in Test Sample Series used for determining TRTC of Styrene in Foods and Food Models*

				Sample Series					
Water	0	0.01	0.02	0.04	0.06	0.08	0.1	0.15	0.2
Emulsions									
3% fat	0	0.06	0.09	0.12	0.15	0.2	0.5	0.7	1
10% fat	0	0.2	0.4	0.8	1.4	2.2	3.2	4.4	5.6
15% fat	0	0.25	0.5	0.75	1	2	3	5	10
20% fat	0	0.25	0.5	1	1.5	2.5	3.5	5	10
25% fat	0	0.25	0.5	1	2	3	5	10	20
30% fat	0	1	2.5	5	7.5	10	15	20	30
Yoghurts									
0.1% fat	0	0.020	0.038	0.043	0.058	0.091	0.096	0.122	0.170
1.5% fat	0	0.031	0.054	0.065	0.080	0.088	0.157	0.250	0.496
3% fat	0	0.057	0.123	0.285	0.452	0.461	0.503	0.516	0.571
Yoghurt drinks (0.1% fat)									
natural	0	0.030	0.044	0.063	0.089	0.121	0.173	0.235	0.349
strawberry	0	0.030	0.039	0.057	0.080	0.114	0.159	0.230	0.335
peach	0	0.030	0.042	0.062	0.090	0.124	0.167	0.234	0.338

Table 3 *TRTC's of Styrene in Water, O/W-emulsions, Yoghurts and Yoghurt Drinks*

	Fat Content (%)	TRTC (ppb)
Water	0	22
Emulsions	3	196
	10	654
	15	1181
	20	1396
	25	1559
	30	2078
Yoghurts	0.1	36
	1.5	99
	3	171
Yoghurt drinks		
natural	0.1	82
strawberry	0.1	92
peach	0.1	94

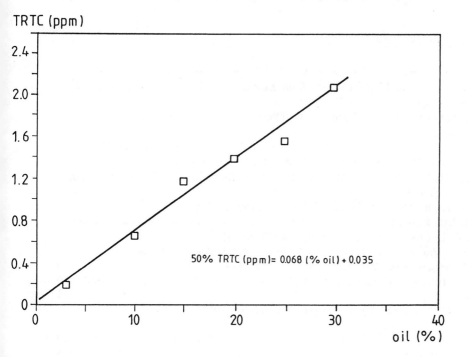

Figure 1 *Recognition threshold concentrations of styrene in O/W-emulsions with different amounts of fat*

Few data about threshold values of styrene are available in the literature. Jensen[9] found a threshold value for styrene in yoghurt at a level of 0.2 ppm, which is in good agreement of the TRTC value for the high fat yoghurt found in this study. Miltz et al.[10] reported a taste threshold value of 5 ppb in sour cream containing 15% fat. The latter value shows a surprisingly low level. However, this value indicates clearly a detection threshold value and the experiments were carried out with a trained panel, which makes the assessors much more sensitive in the detection of styrene. In the present study, an untrained panel was used and a recognition threshold value of 1181 ppb was found for the O/W-emulsion containing 15% fat. Halek and Levinson[11] reported an absolute flavour threshold value for styrene in high fat (26%) cookies of 0.5 ppm, which is also a surprisingly low value. The latter value, however, is weak, because samples with only three different concentrations of styrene were offered to the assessors, resulting in stimulus responses of 0, 63.6 and 100%, respectively. Offering more samples resulting in stimulus responses in between 0 and 100% would be recommended.

4 CONCLUSIONS

TRTC of styrene increases with increasing fat content. In the food models (O/W-emulsions) the TRTC values show a very good linear increase with increasing fat content. The presence of sugar masks the recognition of styrene in yoghurt drinks, while the presence of artificial flavourings influence the TRTC only very slightly. Recogniton of styrene is firstly a function of the fat content of a product.

5 REFERENCES

1. J. Gilbert and J.R. Startin, *J. Sci. Food Agric.*, 1983, **34**, 647.
2. R. Whitney and P.G. Collins, *Bull. Evironm. Contam. Toxicol.*, 1978, **19**, 86.
3. I. Santa Maria, J.D. Carmi, A.G. Ober, *Bull. Environm. Contam. Toxicol.*, 1986, **37**, 207.
4. N. Passy, 'Instrumental Analysis of Foods', G. Charambolous and G. Inglett (eds), Academic Press, New York, 1983, p. 413.
5. J.P.H. Linssen, J.L.G.M. Janssens, J.C.E. Reitsma and J.P.Roozen, *Food Add. Contam.*, 1991, **8**, 1.
6. H. Jenne, *Deutsche Molkerei Zeitung*, 1980, **51/52**, 1906.
7. J.P.H. Linssen, J.L.G.M., Janssens, J.C.E. Reitsma, W.L.P. Bredie and J.P. Roozen, *J. Sci. Food Agric.*, 1993, **61**, 457.
8. S.L. Varner, C.V. Breder and T. Fazio, *J.A.O.A.C.*, 1983, **66**, 1967.
9. F. Jensen, *Ann. Inst. Sup. San.*, 1972, **8**, 443.
10. J. Miltz, C. Elisha and C.H. Mannheim, *J. Food Proc. Preserv.*, 1980, **4**, 281.
11. G.W. Halek and J.J. Levinson, *J. Food Sci.*, 1989, **54**, 173.

A New Method for Determination of Aroma Compound Absorption in Polymers by Accelerator Mass Spectrometry

K. Stenström[1], B. Erlandsson[1], R. Hellborg[1], M. Jägerstad[2], T. Nielsen[2], G. Skog[3], and A. Wiebert[1]

[1]DEPARTMENT OF NUCLEAR PHYSICS, UNIVERSITY OF LUND, SÖLVEGATAN 14, S-22362 LUND, SWEDEN
[2]DEPARTMENT OF APPLIED NUTRITION AND FOOD CHEMISTRY, CHEMICAL CENTER, UNIVERSITY OF LUND, PO BOX 124, S-22100 LUND, SWEDEN
[3]RADIOCARBON DATING LABORATORY, DEPARTMENT OF QUATERNARY GEOLOGY, UNIVERSITY OF LUND, TORNAVÄGEN 13, S-22362 LUND, SWEDEN

1. INTRODUCTION

Accelerator Mass Spectrometry (AMS)[1-3] is a highly sensitive atom-counting method used for detecting very small concentrations of both radionuclides and stable isotopes. Because of the biological importance of carbon, the beta-emitting isotope ^{14}C is so far most frequently studied and used. AMS with ^{14}C is used for dating of geological and archaeological samples, but has also been applied in a variety of other fields, such as oceanography, radioecology and biomedicine. In this report a new application of AMS in food chemistry is reported.

In food chemistry knowledge about the interaction between foodstuffs and packaging material is of importance for development of new packaging systems[4]. Most studies have been made on the migration from the package to the food[5], but during the last few years the interest in studying the absorption of food components into packaging material has increased. Several investigations have shown that flavours can be absorbed to a considerable extent by plastic packages[6-11]. This can lead either to a direct loss of food quality or to damage of the package, which indirectly might affect the foodstuff negatively[11,12]. More knowledge about the absorption of food constituents by packages is still lacking and much work has to be done to understand mechanisms and consequences.

Recently, a project started in Lund exploring the possibilities to use AMS for studying aroma compound absorption in plastic packaging materials[13]. In the AMS method, aroma compounds labelled with ^{14}C are incubated with food packaging polymer films. After converting the polymer samples to elementary carbon, the isotopic composition of the samples is measured by the AMS facility[14] at the Lund Pelletron tandem accelerator and the degree of aroma compound absorption is revealed. The high sensitivity of AMS, gained from atom counting, makes it possible to use both small pieces of packaging material and low concentrations of ^{14}C, yet requiring short measuring times.

At the Department of Applied Nutrition and Food Chemistry in Lund a new method using supercritical fluid extraction (SFE) coupled directly to gas chromatography (GC) has been developed for extracting and analysing aroma compounds absorbed into polymer films[15-17]. It compares favourably with previously used techniques, especially in terms of time, labour and sample size. However, with the current instrumentation the SFE-GC method has some limitations, e. g. it is not possible to extract polar substances with the SFE-equipment available, or to quantify non-volatiles by GC. Sorption of certain compounds can be studied with both techniques, and comparison of the recoveries of the two methods can be made. After optimising the AMS-technique, it can provide an important tool for studying sorption of analytes not possible to determine by SFE-GC.

2. GENERAL DESCRIPTION OF AMS

As can be concluded from the name accelerator mass spectrometry the method is an extension of ordinary mass spectrometry by including an accelerator, preferably an electrostatic tandem accelerator (Figure 1). In ordinary mass spectrometry isotopic concentrations below about 10^{-7} relative to the most common isotope can *not* be measured due to inseparable interference from isobars.

The following properties, combined with electrostatic and magnetic spectrometers, are among the most essential for enabling an AMS-system to measure isotopic ratios down to 10^{-15}:

1) Interference from isobars can be avoided in many cases by using a *negative ion source*. For instance, ^{14}C analysis is enabled due to the fact that the isotope ^{14}N is eliminated because of the instability of negative nitrogen ions, while negative carbon ions are easily produced.

2) *The stripping process* at the accelerator high voltage terminal is of great importance since it breaks up molecules through the removal of three or more electrons. By selection of a high charge state with the high-energy analysing magnet, molecular isobaric interferences are thus removed. Molecular isobars such as ^{13}CH and $^{12}CH_2$ are eliminated in this way.

3) *The high final ion energy*, tens of MeV, provides such a good resolution with energy or energy-loss measurements, that every single ion can be identified in atomic and mass numbers.

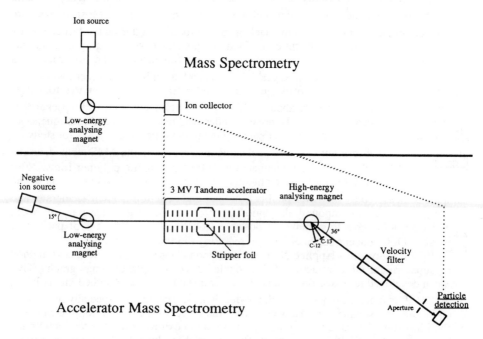

Figure 1. *Comparison of mass spectrometry and accelerator mass spectrometry*

3. EXPERIMENTAL PROCEDURE

An aqueous solution with 1 ppm of the ^{14}C-labelled aroma compound is prepared. Strips of 10-20 mg of the polymer film is stored in 5 ml of the aqueous solution in glass ampoules, sealed by flame. The ampoules are stored in the dark at 4°C for one week[16].

To do AMS-measurements the polymer films with absorbed aroma compounds have to be converted to elementary carbon. After removing the polymer film from the solution, the polymer is wiped dry and the plastic sample is combusted to carbon dioxide with CuO in a sealed quartz tube. Thus, all carbon in the polymer as well as in the absorbed aroma compounds is converted to carbon dioxide. The reduction method used in Lund is based on the production of elemental carbon by the catalytic reduction of carbon dioxide over an iron-group metal powder[18]. Our sample preparation system (Figure 2) has previously been used successfully mainly for carbonate samples[19-21] but also for organic materials such as charcoal. Combustion of processed plastic samples is somewhat different. It is of extreme importance that the combustion to carbon dioxide of both the polymer and the aroma compounds is complete, a demand which has been shown to be fulfilled with sealed-tube combustion. The use of ethyl acetate involves combustion of a liquid, different from combustion of dry matter, for which closed-tube combustion is not necessary. Using closed tubes, the whole combustion volume can be heated, which prevents liquids from condensing on cooler parts of the system, and thus escape combustion. The reduction step takes 1-4 hours.

The prepared carbon samples are analysed using the AMS system at the 3UDH Pelletron tandem accelerator in Lund[14,22]. By measuring the isotopic composition of the samples the degree of aroma compound absorption is revealed. Each sample requires about 20 minutes of measuring time.

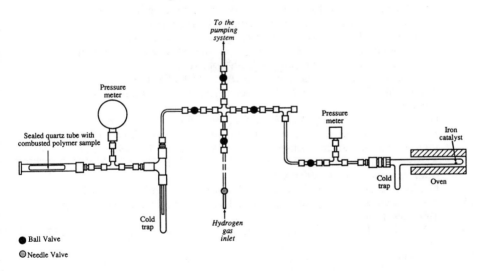

Figure 2. *Outline of the sample preparation system*

4. RESULTS AND DISCUSSION

The background in the AMS system is at the moment about 3 % of the NBS oxalic acid standard[23], and samples of untreated polyethylene have a ^{14}C content of about 10% of the standard. The reason for the relatively high background of the untreated polyethylene samples could be absorption of atmospheric carbon dioxide into the polymer. The ^{14}C-labelling of the ethyl acetate is therefore adapted to give polymer samples with activities of two orders of magnitude higher than the polyethylene background. The measuring error in the accelerator is presently about ±5%.

Preliminary studies of the sorption of ethyl acetate by low density polyethylene are in progress. Further studies of fatty acid sorption by polyethylene are also planned.

Since the elemental carbon production and the AMS measurement are established, well-tested and the same as for other type of samples, the AMS technique should be suitable for studying the absorption of aroma compounds into plastic packaging materials. In the near future we will evaluate the SFE-GC method and apply   AMS on components that cannot be studied by SFE-GC with its current instrumentation.

References

1. W. Kutschera, *Nuclear Physics News*, 1993, **3(1)**, 15
2. A.E. Litherland, *Nucl. Instr. and Meth.,* 1984, **B5**, 100
3. W. Wölfli, *Europhysics News,* 1984, **15(2),** 1
4. J.H. Hotchkiss, 'Food and Packaging Interactions', American Chemical Society, Washington, DC, 1988, p.1
5. S.J. Risch, *Food Technol.,* 1988, **7** ,95
6. C.H. Mannheim, J. Miltz, N. Passy, 'Food and Packaging Interactions', American Chemical Society, Washington, DC, 1988, p.68
7. G.W. Halek, M.A. Meyers, *Pack. Technol. Sci.,* 1989, **2**, 141
8. A.P. Hansen, D.K. Arora, 'Barrier Polymers and Structures', American Chemical Society, Washington , DC, 1990, p.318
9. T. Imai, B.R. Harte, J.R. Giacin, *J. Food Sci.,* 1990, **55**, 158
10. T. Ikegami, K. Nagashima, M. Shimoda, Y. Tanaka, Y. Osajima, *J. Food Sci.,* 1991, **56**, 500
11. G.D. Sadler, R.J. Braddock, *J. Food Sci.,* 1991, **56**, 587
12. K. Hirose, B.R. Harte, J.R. Giacin, J. Miltz, C. Stine, 'Food and Packaging Interactions', American Chemical Society, Washington, DC, 1988, p.28
13. K. Stenström, B. Erlandsson, R. Hellborg, A. Wiebert, G. Skog, T. Nielsen, *Nucl. Instr. and Meth.,* 1994, **B89**, 256
14. G. Skog, R. Hellborg, B. Erlandsson, *Radiocarbon* , 1992, **34(3)**, 468
15. T.J. Nielsen, I.M. Jägerstad, R.E. Öste, B.T.G. Sivik, *J. Agric. Food Chem.,* 1991, **39**, 1234
16. T.J. Nielsen, I.M. Jägerstad, R.E. Öste, B.O. Wesslen, *J. Food Sci.,* 1992, **57**, 490
17. T.J. Nielsen, I.M. Jägerstad, R.E. Öste, *J. Sci. Food Agric. ,* 1992, **60**, 377
18. J.S. Vogel, J.R. Southon, D.E. Nelson, T.A. Brown, *Nucl. Instr. and Meth.,* 1984, **B5**, 289
19. K. Stenström, B. Erlandsson, R.Hellborg, K. Håkansson, A. Wiebert, G. Skog, Internal Report LUNFD6/(NFFR-3061)/1-31/(1993)
20. K. Stenström, B. Erlandsson, R.Hellborg, K. Håkansson, A. Wiebert, G. Skog, Internal Report LUNFD6/(NFFR-3062)/1-27/(1993)

21. K. Stenström, S. Leide-Svegborn, B. Erlandsson, R.Hellborg, S. Mattson, L-E. Nilsson, B. Nosslin, G. Skog, A. Wiebert, Internal Report LUNFD6/(NFFR-3063)/1-14/(1993)

22. R. Hellborg, K. Håkansson, G. Skog, *Nucl. Instr. and Meth.*, 1990, **A287**, 161

23. I. Karlén, I.U. Olsson, P. Kållberg, S. Kilicci, *Arkiv för Geofysik*, 1964, **4(22)**, 465

Mechanical Interactions/Barrier Properties

Compatibility of Plastic Materials with Foodstuffs: Mechanistic and Safety Aspects of Ionized Polypropylene

A. Feigenbaum[1], D. Marqué[2], and A-M. Riquet[2]

[1]INSTITUT NATIONAL DE LA RECHERCHE AGRONOMIQUE, 23, RUE CLÉMENT ADER, 51100 REIMS, FRANCE
[2]INSTITUT NATIONAL DE LA RECHERCHE AGRONOMIQUE, LNSA, 78352 JOUY-EN-JOSAS, FRANCE

1 INTRODUCTION

The compatibility of plastic packaging materials with foodstuffs is largely concerned with migration of constituents of the plastic to food. Two kinds of contaminants have to be considered :
- Substances whose presence can be predicted, because they are on positive lists of national or European authorities : these are mainly monomers and additives used to manufacture plastics.
- Substances whose presence cannot be predicted, and which do not belong to the positive lists :
    - Contaminants linked to recycling of post consumer wasted plastics : these may be any chemical,
    - Substances which have a link to those on positive lists, because they are reaction products of monomers, of polymers or of additives.

We will focus here on the latter category in the case of polypropylene (PP) or of polyolefins containing propene as a co-monomer. To have an overview of the reactivity in PP, we should first start with the reactivity of the polymer itself. PP is such an unstable polymer, that in the time of its discovery, it was thought that it could never be a commercial material[1].

$$PP\text{-}H \text{ ---> } PP° \qquad \text{(Equation 1)}$$
$$PP° + O_2 \text{ ---> } PP\text{-}OO° \qquad \text{(Equation 2)}$$
$$PP\text{-}OO° + PP\text{-}H \text{ ---> } PP\text{-}OOH + PP° \qquad \text{(Equation 3)}$$

When submitted to heat, light or ionizing treatments, PP (PP-H) generates alkyl radicals PP°, which are reducing species (Equation 1). These alkyl radicals are rapidly trapped by oxygen to give peroxyl radicals PP-OO° (Equation 2). The peroxyl radicals are strong oxidants, which attack a new molecule of PP (Equation 3). Reaction (3) represents a chain propagation, which induces a great damage in the polymer : on one hand a new alkyl radical is produced which reacts immediately with oxygen (Equation 2) to produce a new peroxyl radical. On the other hand, reaction (3) leads to a hydroperoxide PP-OOH, which leads to new oxidizing radicals (PPO°, OH°), to chain cleavage, and to the formation of hydroxyl, carbonyl or carboxyl functional groups.

To mind these reactions, stabilizers are used. For instance a phenolic antioxidant reacts with the peroxyl radical and replaces PP-H in the hydrogen abstraction steps (Equations 1 and 3) : the aryloxy radical ArO° thus formed (Equation 4) is highly stabilized both by electron delocalization and by sterical shielding. Other antioxidants like the aromatic phosphites are intended to destroy the hydroperoxides (Equation 5) :

$$PP\text{-}OO° + ArOH \longrightarrow PP\text{-}OOH + ArO° \qquad \text{(Equation 4)}$$
$$(ArO)_3P + PP\text{-}OOH \longrightarrow (ArO)_3P{=}O + PP\text{-}OH \qquad \text{(Equation 5)}$$

Clearly, if knowledge is needed about reaction products formed in polymers, then one should start by studying the behaviour of stabilizers, which are intended to react.

2 ANALYTICAL STRATEGIES

Analytical strategies which allow to determine whether migration stays within safe limits start with an analysis of extracts of the material. In many cases, further testing is not necessary[2]. The available analytical methods can be classified according to the molecular weight of the migrants, which is an approximate but useful approach.

Since the EC Practical Guide of the Commission[3] states that substances with a molecular weight higher than 1000 g x mol^{-1} are not likely to be absorbed by the gastrointestinal tract, we should focus on substances with $M \leq 1000$ g.mol^{-1}.

The Dutch enforcement method[4] involves an extraction with ether followed by GC analysis. However substances with $M > 700$ g.mol^{-1} are not eluted, whereas peaks of solutes with $M < 90$ g.mol^{-1} are masked by the solvent.

It is possible to evaporate the solvent and to proceed to HPLC, NMR or IR. These techniques are not limited on the high molecular weight side, but substances with $M < 250$ g.mol^{-1} are evaporated along with the solvent.

Other methods, like superfluid extraction coupled to GC or SFC[5] give reliable results for $M > 200$ g.mol^{-1}.

The volatiles, which have a high migration rate, are not satisfactorily analysed by all these methods. They require appropriate methods, based mainly on headspace techniques[6].

Clearly, several complementary methods have to be used in order to analyse the complete scope of the potential contaminants. We have suggested that group methods, based on the functional classes, could be very useful to control the migration from packaging : plasticizers[7], aromatic antioxidants[8] or hindered amine stabilizers[9]. A functional class is a group of substances having the same functional group, which can be detected by a spectroscopic method. The functional group can also be responsible of similarities in the technological functions and, sometimes in the metabolic behaviour of these contaminants.

3 AROMATIC ANTIOXIDANTS AS A FUNCTIONAL CLASS

3.1 Phenolic Antioxidants

A typical reactivity of phenolic antioxidants in PP exposed to heat light or radiations is shown in Scheme 1, with compound I, which is a common stabilizer. Successive H abstractions by PP-OO° (Equation 4) give quinonoïd type structures (II, IV and VI). II and IV isomerize to new phenolic type structures (III and V), which exhibit also antioxidant activity. I and its derivatives react successively with four peroxyl radicals. The final products is the dimer VI, with a highly conjugated structure, responsible of discoloration of the materials[10,11].

It has been reported that some products are formed specifically during ionization processes[12], which may help to demonstrate whether ionization treatments have been applied to packaging materials.

3.2 Aromatic Phosphites

Phosphites are often used to destroy hydroperoxides (Equation 5). They usually possess an aryl group which improves their resistance to hydrolysis[13]. The phosphite VII is commonly encountered along with VIII (Equation 5), but also with IX and X[14] (Scheme 2).

3.3 HNMR as an Analytical Tool for Exploration of PP Extracts

Both phenolic and phosphite type antioxidants have aromatic rings which can be characterized by spectroscopic methods : in UV by their strong P → P* transition band around 260 nm, and in HNMR by the aromatic protons around 7 ppm chemical shift. Figure 1 shows the spectrum of extracts of polyolefin granules, before (Figure 1A) and after (Figure 1B) irradiation with an electron beam (80 kGy). It can be seen that after irradiation, the aromatic stabilizers VII and XV [XV : pentaerythritol tetrakis[3-(3,5-ditert.butyl-4-hydroxyphenyl)propionate]] have been replaced by new compounds. HNMR thus appears as a quick method which provides information on the possible presence or absence of potential contaminants.

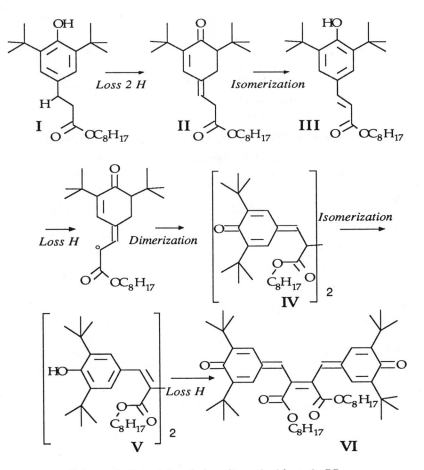

Scheme 1 *Reactivity of phenolic antioxidants in PP*

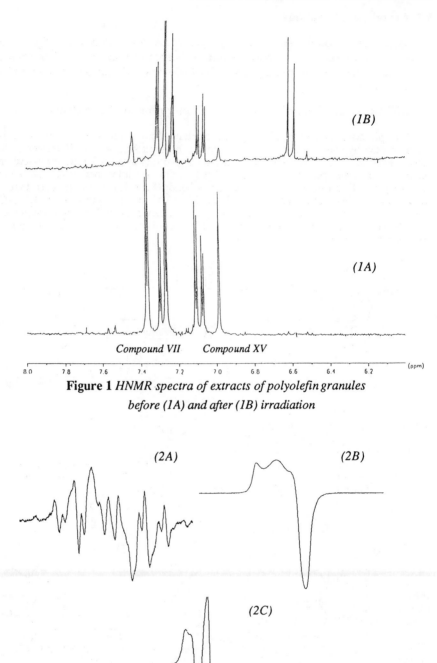

Figure 1 *HNMR spectra of extracts of polyolefin granules before (1A) and after (1B) irradiation*

Figure 2 *ESR spectra of the radicals observed in PP after irradiation PP° (2A), PP-OO° (2B), aminoxyl radical (2C)*

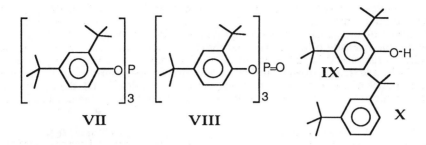

Scheme 2 *Derivative forms of phosphite stabilizers*

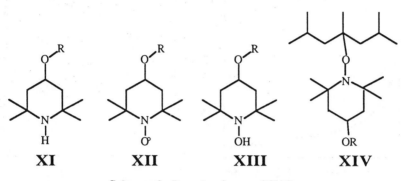

Scheme 3 *Reactive forms of HAS*

4 REACTIVITY OF HINDERED AMINE STABILIZERS (HAS)

4.1 Decay of Peroxyl Radicals

Hindered amine stabilizers (HAS) like XI can exist under different forms, which can all be reactive : the aminoxyl radical XII, the hydroxylamine XIII and the adduct XIV (Scheme 3).

XI + PP-OOH ---> XIII + ROH	secondary antioxidant (Equation 6)
XII + PP° ---> XIV	radical coupling (Equation 7)
XII + PP° ---> PP(-H) (alkene) + XIII	hydrogen transfer (Equation 7)
XIII + PP-OO° ---> XII + PP-OOH	primary antioxidant (Equation 8)
XIII + 1/2 PP-OOH ---> XII + 1/2 PP-OH	(Equation 9)

HAS alone are not very efficient stabilizers, but associated to a phenolic antioxidant they dramatically improve the stability of PP, specially during irradiation[15]. Using electron spin resonance (esr) radicals are specifically detected, and results will be presented here with peroxyl and with aminoxy radicals.

We have exposed a PP film, stabilized with 0.05 % XV and 0.3 % HAS XI (Tinuvin 770, Ciba-Geigy) to an electron beam (10, 20 and 40 kGy). Before ionization, almost no aminoxyl radical XII could be detected. Two kinds of experiments were made : films were irradiated either in air (Experiments A) or under vacuum (Experiments V), in sealed esr tubes, at room temperature. The tubes irradiated under vacuum were opened in air either 1 h (Exp. V-1h) or 10 d (Exp. V-10d) after ionization.

The samples A showed the spectra of PP-OO° radicals, which were stable for months.

In samples V, only the characteristic spectrum of PP° radicals (Figure 2A) could be observed after ionization (Equation 1). Opening the tubes in air resulted in a rapid appearance of the signal of the peroxyl radicals PP-OO° (Figure 2B). After a few minutes, the intensity of the signal went through a maximum and started to decrease rapidly. In samples V-1h, the decay of the peroxyl radicals was very rapid during the first 24 h, and much slower later. It took 4 weeks to achieve almost complete disappearance of the peroxyl radicals (Figure 3). In samples V-10d, the decay of the peroxyl radicals was much faster, and their signal completely disappeared within 24 h.

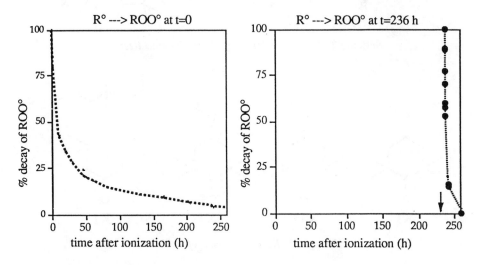

Figure 3 *Decay of peroxyl radicals*

It thus appears that the later the oxygenation, the faster the decay of the peroxyl radicals formed. This indicates that under vacuum a part of the alkyl radicals reacts rapidly, by interchain reactions. This can only happen in condensed zones of the polymer (Scheme 4).

PP° + PP° ---> PP-PP (cross linking) or PP(-H) (alkene)

When these reactive PP° are trapped by O_2 (samples A, V-1h), they give rise to peroxyl radicals which decay slowly. This differentiated behaviour of radicals in PP has been attributed to differences in their mobility[16]. An alternative explanation is that peroxyl radicals in amorphous zones can be reached and destroyed by the stabilizers (Equations 3 and 8). The slowly decaying peroxyl radicals observed would then be localized in condensed zones of the material, where stabilizers do not diffuse. They react by sequences of (reaction 3 + reaction 2), which leaves the overall number of PP-OO° constant and explains the slow decay.

4.2 Aminoxyl Radicals

The preceeding rationalization could be supported if aminoxyl radicals could be formed at early times after exposure in air. Unfortunately, their spectrum is masked by that of the peroxyl radicals. They can only be observed in the film when the PP-OO° have significantly decayed, after several weeks in air (samples V-10d) (Figure 2C).

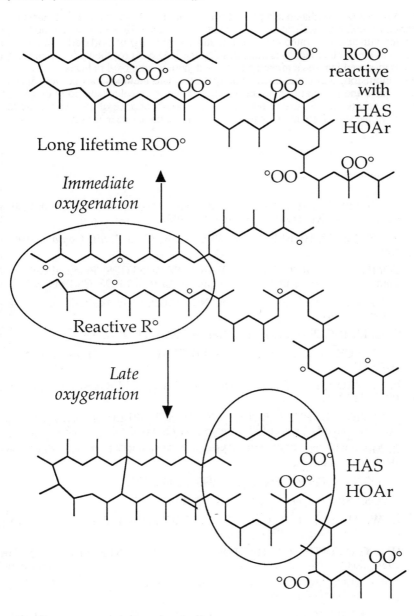

Scheme 4 *Post-irradiation oxidation of PP*

We thought that aminoxyl radicals,which are low molecular weight compounds, could be detected after selective migration into a liquid in contact. When the irradiated PP (vacuum 1 h) was left 10 days at 50°C in contact with methyl palmitate, the aminoxyls were detected in the liquid, which allows them to be recognize , even when they cannot be seen in the film.

When a non irradiated sample of film was put in contact with methyl palmitate, no migration of aminoxyl radicals was detected, as expected. However, when the film was in contact with an unsaturated fatty ester, methyl oleate, largely oxidized, a large amount of $NO°$ was again obtained. This indicates that the HALS XI (or the hydroxylamine XIII) itself has migrated into the simulant and reacted there to aminoxyl radicals.

In conclusion, we have shown that spectroscopic methods can be very useful to evaluate food and packaging interactions, both from the mechanistic point of view, and in order to assess the safe use of the materials. The importance of the knowledge of the reactivity of the stabilizers in the material and in the food has been illustrated.

References

1. N. GRASSIE and G. SCOTT, 'Polymer Degradation and Stabilisation', Cambridge University Press, Cambridge, 1988, Chapter 1, p. 1.

2. R. FRANZ, M. HUBER and O. PIRINGER, *Food Addit. Contam.*, 1994, in press.

3. COMMISSION OF THE EUROPEAN COMMUNITIES, 'Practical Guide N°1', Commission of the European Communities, Brussels, 1993, CS/PM/2024.

4. D. VAN BATTUM and J. B. H. VAN LIEROP, *Food Addit. Contam.*, 1988, **5** (Suppl. 1), 381.

5. O. G. PIRINGER, *Food Addit. Contam.*, 1994, **11** (2), 221.

6. J. LE SECH, V. DUCRUET and A. FEIGENBAUM, *J. Chromatogr. A*, 1994, **667**, 340.

7. E. MONROY, N. WOLFF, V. DUCRUET and A. FEIGENBAUM, *Analusis*, 1993, **21**, 221.

8. J. EHRET-HENRY, J. BOUQUANT,D. SCHOLLER, R. KLINCK and A. FEIGENBAUM, *Food Addit. Contam.*, 1992, **9** (4), 303.

9. D. MARQUÉ, A. FEIGENBAUM and A-M. RIQUET, *Polym. Engineering*, 1994, to be published.

10. J. POSPISIL, *Polym. Deg. Stab.*, 1993, **39** (1), 103.

11. J. POSPISIL, *Polym. Deg. Stab.*, 1993, **40** (2), 217.

12. D. W. ALLEN, M. R. CLENCH, A. CROWSON and D. A. LEATHARD, *Polym. Deg. Stab.*, 1993, **39** (3), 293.

13. H. R. GAMRATH, R. E. HATTON and D. E. WEESNER, *Indust. Eng. Chem.*, 1954, 208.

14. Z. EL MAKHZOUMI, Thèse de Docteur ès sciences (chimie industrielle), Université de Reims, 1991.

15. J.F. RABEK, 'Photostabilization of polymers : principles and applications', Elsevier Appl. Sci., London, 595 p.

16. D. J. CARLSSON, C. J. B. DOBBIN and D. M. WILES, *Macromol.*, 1985, **18** (9), 1791.

Interaction Phenomena in Multilayer Flexible Packaging

G. Pieper

TETRA PAK (RESEARCH) GMBH, WALDBURGSTRASSE 79, D-70563 STUTTGART, GERMANY

1 INTRODUCTION

The protection of food is a key function of a package. It is therefore vital for the quality of the packed product that the packaging material assures the protection of the food from undesirable environmental influences over the whole shelf life. The permeation of odour and/or oxygen through the packaging material into food, the permeation/ absorption of food components into the packaging material as well as the migration of packaging material components into food can have detrimental effects on the product quality. However, interaction phenomena between the packaging material and the food based on permeation/absorption may have as a secondary effect an impact on the functionality and the integrity of the package. This paper focuses on such interaction phenomena in multilayer packages with primarily aluminium foil as a barrier layer and LDPE as product contact layer.

2 GENERAL CONSIDERATIONS

The consequences of interaction phenomena between the packaging material and the food are generally different for packaging materials with and without barriers like aluminium foil. In packaging materials without a barrier layer, the main interaction effect is based on the permeation of food components through the packaging material. The extent of permeation depends on the solubility and the diffusivity of the respective food component in the packaging material. A permanent loss of food components may be observed, whereas the initiation of package defects is of minor importance. Interaction in a packaging material with a barrier interlayer can be described on the basis of absorption/permeation of food components into the food contact polymer layer. Absorption into packaging materials with barrier layers leads to an equilibrium level. The solubility of the food components in the packaging material and the respective partition coefficients assign how much of a compound is absorbed in this stage. The interaction between food and multilayer packages with barrier layers has shown a variety of possible impacts on the functionality and the integrity of packages, which subsequently may lead to undesirable changes in the food quality (Figure 1).

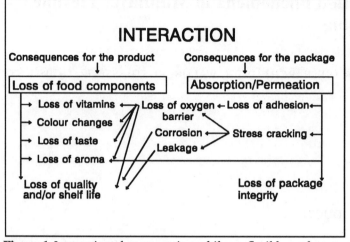

Figure 1 *Interaction phenomena in multilayer flexible packages and their consequences for the product.*

3 ADHESION

Absorption and permeation of acid food components into the food contact polymer layer and in a later stage the accumulation of these compounds at the interface between the polymer tie resin layer and the aluminium foil can lead to adhesion losses or complete delamination. The adhesion between the aluminium foil and the polyethylene in multilayer flexible packages is preferably achieved by the use of copolymer tie resins containing carboxyl groups. These acid groups interact with the basic sites on the aluminium foil and bonds are formed between the two layers. Acid food components, which are able to permeate through polyolefins and then compete with the carboxyl

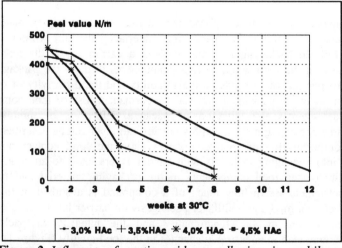

Figure 2 *Influence of acetic acid on adhesion in multilayer flexible packages with aluminium foil.* [1]

groups of the tie resin may therefore initiate delamination. Acetic acid is one example of such a food acid.

Acetic acid solutions (Figure 2) in concentrations of 3-4.5% were stored in contact with multilayer structures with LDPE and aluminium foil for several weeks at 30°C. A clear correlation was found between the decrease in the peel value (which is a measure of adhesion in multilayer packages) and the acetic acid concentration.

Delamination in multilayer flexible packages with aluminium foil will subsequently lead to a loss of oxygen barrier properties with detrimental effects on the product quality (Figure 1). It is therefore necessary that good adhesion is assured over the whole shelf life of the product. To improve the adhesion durability in multilayer packages, different approaches are possible. One possibility is to increase the thickness of the copolymer tie resin layer.

Figure 3 shows the influence of the PE-copolymer thickness on the adhesion durability in multilayer packages filled with orange juice. The thicker the adhesive layer, the higher is the resistance against delamination effects of orange juice. Another possibility to achieve adhesion durability is the use of an additional barrier against permeation of food acids in the internal polymer layers. Different polymers were tested in this respect and Figure 4 shows the barrier properties of these polymers against the permeation of acetic acids from orange juice as a food simulant. (Orange juice was used as a matrix because the permeation of food components through polymers can be influenced by swelling effects, e.g. due to absorption of aroma compounds.)

Polyester showed the best barrier properties against acetic acid permeation, followed by LDPE/EAA and PA. EVOH performed surprisingly poorly when compared with its oxygen and aroma barrier properties. This indicates that one has to be cautious in generally concluding from other barrier properties on absorption/permeation barrier properties. Based on these results, an improvement of the adhesion durability in aluminium foil/LDPE structures can be achieved with an additional PET-layer.

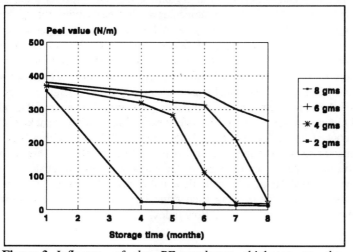

Figure 3 *Influence of the PE-copolymer thickness on the adhesion durability. Packages filled with orange juice and stored at 30°C.* 2

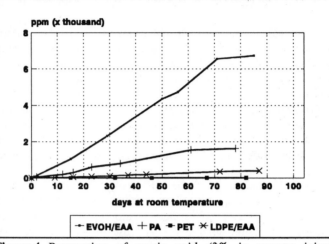

Figure 4 *Permeation of acetic acid (3% in orange juice) through polymers.*

4 ENVIRONMENTAL STRESS CRACKING OF POLYETHYLENE

Polyethylene is known to be sensitive to environmental stress cracking (ESC), when in contact with fats and oils. The polymer gets brittle and cracks, especially at points of high mechanical stress.

In the following, an explanation of the mechanism of ESC by Lustiger & Markham is summarized.[3] They discussed the mechanism of the brittle failure of polyethylene as a result of lower stress levels over a long period of time. These low level stresses affect amorphous polymer chains between the crystalline regions, especially the tie molecules between adjacent lamellae. These tie molecules are not continuous chains, but consist of many entanglements, which under long term low level stress begin to untangle until finally the material fails. This type of failure will be accelerated in the presence of lubricants or plasticizers. Fats and oils and, surprisingly, also formulated tomato products were found to be effective in this respect. Stress cracking together with salt and acids as constituents, e.g. in tomato formulations, present a high risk of subsequent corrosion of the aluminium foil in multilayer packages. However, fats, oils, high fat sauces and tomato products can still be safely packed in multilayer packages with LDPE as product contact layer. The key point is the use of a PE grade with a higher resistance to environmental stress cracking. According to Lustiger et al., [3,4] higher resistance to ESC will be achieved by an increase in the amount of tie molecules. PE with high molecular weight and hence low melt flow index provides more tie molecules and is therefore more suitable for packing such critical products, which are known to cause ESC in ordinary PE with a high melt flow index.

5 AROMA ABSORPTION

The absorption of citrus aroma compounds into multilayer packages is a matter of importance, even if it does not show obvious effects on the package integrity. Based on the general equation that permeation is a product of diffusivity and solubility, only compounds with high diffusion coefficients and high solubility coefficients in the polymer implement the risk of extensive losses in the product. For polyolefins, unpolar aroma compounds like limonene, the major aroma compound in citrus juices, are the most critical ones in regard to solubility following the rule that like dissolves like. While there are minor differences in solubility, the diffusion coefficient and consequently permeation rates differ by order of magnitudes decreasing from LDPE > HDPE > PP. Polar polymers like PET show basically very low diffusion coefficients for polar and unpolar aroma compounds respectively, which results in good barrier properties for both.

Most multilayer cardboard packages for citrus juices contain LDPE as product contact layer and hence show absorption mainly of unpolar aroma compounds of citrus juices. The extent of absorption in packages with a barrier layer depends on the thickness of the internal product contact layer(s) and the area/volume ratio of the packages.[5] Its impact on the product quality has been discussed extensively - often on the basis of analytical evaluation of model systems. However, the correlation between analytical results and the actual sensory performance of the citrus juices has seldom been investigated. The following discusses results regarding the effect of different levels of citrus flavour scalping on the sensory performance of high quality orange juice during 23 weeks at 4°C.[6]

Different levels of flavour scalping were achieved by the use of package types with different PE/copolymer tie resin thicknesses (between 13 and 42 gms/m[2]) and different area/volume ratios. Figure 5 shows the limonene retention of orange juice, stored in these packages compared with a glass reference. Since the packaging materials contained efficient barrier interlayers an equilibrium level was reached after ≈ 2 weeks.

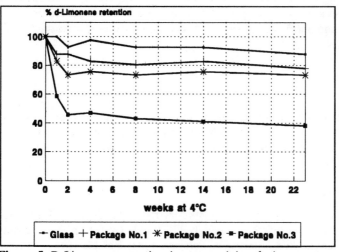

Figure 5 *D-Limonene retention in orange juice during storage at 4°C in different packages. The value found in glass bottles after 1 wk of storage (0.020 V v/v d-limonene) was taken as initial concentration of d-limonene assuming that no chemical degradation has taken place within 1 wk at 4°C.*
Reprinted from J. Food Sci.[6]

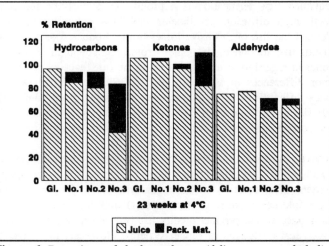

Figure 6 *Retention of hydrocarbons (d-limonene excluded), ketones, and aldehydes in orange juice and recovered hydrocarbons (d-limonene excluded), ketones, and aldehydes from packaging material extracts after 23 wk storage in different packages at 4°C. Reprinted from J. Food Sci.* 6

18 other orange juice aroma compounds were monitored during storage of which mainly hydrocarbons, ketones and aldehydes were more or less affected by the absorption into the PE-inside coating (Figure 6). However, an experienced panel did not distinguish between orange juice stored in glass bottles and that stored in the laminated cardboard packages. An absorption of up to 50% limonene and other hydrocarbons, small amounts of ketones, and aldehydes had no significant influence on the sensory quality of orange juice stored at 4°C.

6 CONCLUSION

In order to assure the protection of the food by the package until the end of the required   shelf life, the specific interaction phenomena have to be understood and have to be taken into consideration when the most suitable packaging material for the specific food is selected.

References

1. K. Petersén, E. Johansson, Tetra Pak Materials AB, Lund, personal communication.
2. K. Henningsson, Tetra Pak Materials AB, Lund, Sweden, personal communication.
3. A. Lustiger and R. L. Markham, *Polymer*, 1983, **Vol 24**, 1647.
4. A. Lustiger, 'Engineered Material Handbook', ASM Handbook Committee, USA, 1988, Vol. 2, 796.
5. C.E. Sizer, P.L. Waugh, S. Edstam, and P. Ackermann, *Food Technol.*,1988, **42 (6)**, 152.
6. G. Pieper, L. Borgudd, P. Ackermann and P. Fellers, *J. Food Sci.*, 1992, **57 (6)**, 1408.

Interactions between Carboxylic Acids and Laminated Food Packaging Material

G. Olafsson[1] and I. Hildingsson[2]

[1]DEPARTMENT OF APPLIED NUTRITION AND FOOD CHEMISTRY, CHEMICAL CENTER, LUND UNIVERSITY S-22100 LUND, SWEDEN
[2]DEPARTMENT OF CHEMICAL ENGINEERING, CHEMICAL CENTER, LUND UNIVERSITY, S-22100 LUND, SWEDEN

1 INTRODUCTION

In the food packaging industry, it is well known that certain food products can cause adhesion problems in laminated food packaging material. In the worst case, a delamination may occur, which affects the functionality of the package and possibly the quality of the food[1]. Among foods that are difficult to pack are acidic foods and foods with a high fat content.

The purpose of this work was to study the effect of some common organic acids and fatty acids on the adhesion between low density polyethylene (LDPE) and aluminium foil (Al foil), and furthermore to investigate possible synergistic effects between acetic acid and fatty acids.

2 MATERIALS AND METHODS

2.1 Food Packaging Material

The packaging material used in this work was a flexible laminate, similar to that used for the aseptic packaging of milk. The material had the following composition:
LDPE($15g/m^2$) / paper($186g/m^2$) / LDPE($25g/m^2$) / Al-foil(6.5μm) / LDPE($15g/m^2$) / LDPE($25g/m^2$). No tie layers were used, but the LDPE layer between the Al foil and the innermost layer was ozone-treated. The LDPE (Exxon LD 256) had a density of 0.92 g/cm^3.

2.2 Design of the Study

Five different studies were made, two with organic acids, two with fatty acids and one with both fatty acids and acetic acid. Table 1 gives an overview of the carboxylic acids used, and also the model systems and experimental conditions used in each study.

2.3 Test Packaging

Envelopes of approximately 10 x 20 cm in size were produced in a thermo-sealer. The envelopes were filled with 45 ml of the test solution and stored at room temperature.

Table 1 Experimental Design

Study	Carboxylic acid	Conc. % (wt/wt)	Solvent / blank	Time of analysis (days)	Methods
I[5]	Acetic	0.5, 1, 3	water	1, 2, 3 , 4, 7, 14, 21, 28	Peel test FTIR (ATR) ESCA
II[6]	Acetic Propionic Citric Lactic	equivalent* equivalent equivalent	water	1, 2, 3, 4, 5, 7, 14, 21, 28	Peel test FTIR (ATR/RAS) ESCA
III[7]	$C_{16:1}$ $C_{18:1}$ $C_{18:2}$ $C_{18:3}$	0.2 0.2 0.2 0.2	water/eth./SDS (87:10:3)	1, 2, 3, 4, 7, 10, 14, 21	Peel test Sorption FTIR (ATR) Contact angle
IV[7]	$C_{10:0}$ $C_{12:0}$ $C_{14:0}$ $C_{16:0}$ $C_{18:0}$ $C_{18:1}$ $C_{18:2}$	1 1 1 1 1 1 1 1	ethanol (95 % wt/wt)	1, 7, 14, 21, 28, 56	Peel test Sorption FTIR (ATR) Contact angle
V[8]	$C_{18:1}$ Acetic acid $C_{18:1}$ + Acetic acid	1 2,7 1 2,7	water / rapeseed-oil / Tween-80 (90:9.7:3)	1, 2, 4, 7, 14, 21, 28	Peel test Sorption FTIR (ATR/ RAS) Contact angle

* equal number of meq/ml as in the acetic acid solution

2.4 Methods

The following measurements were made: (i) the adhesion between the inner LDPE layer and the Al foil, using a 180° peel test, (ii) the sorption of free fatty acids by LDPE, (iii) the presence of carboxylic acids on the inner LDPE film was estimated by FTIR using the attenuated total reflection technique (ATR), (iv) the presence of carboxylic acids on the Al foil was estimated by FTIR using reflection absorption spectroscopy (RAS) and ESCA, (v) the wettability of the Al foil was estimated using contact angle measurements and (vi) the permeation of carboxylic acids through the LDPE film was measured using special permeation cells.

3 RESULTS AND DISCUSSION

Listed below are the most important results of studies I-V. The results will be described for each acid or group of acids, independent of in which order the studies were made, beginning with the organic acids.

3.1 Organic Acids

3.1.1 Acetic Acid. A 3% aqueous solution of acetic acid (Study II) was shown to have a strong effect on the adhesion between LDPE and Al foil (Figure 1). After 3 days, the two layers were totally delaminated and remained so for 4 days, after which the adhesion started to recover and had reached approximately 50% of its initial value on day 7, after which it remained unchanged during the study. A similar adhesion behaviour was observed in separate experiments with acetic acid in water (Study I) and in an oil in water emulsion (Study V). In all three experiments a delamination occurred in 3-4 days, followed by a recovery after 7-14 days of storage at room temperature. At lower acid concentrations the effect on adhesion was much less. The material in 1% acetic acid gradually lost its adhesion during storage and after 4 weeks the adhesion was ~ 80% of its original value. Neither the material in 0.5% acetic acid nor the reference sample in water were significantly affected during the storage time.

The delamination is believed to be caused by the permeation of the acetic acid. An aliphatic carboxylic acid such as acetic acid is slightly polar, but is believed to dissolve in the non-polar LDPE matrix as a dimer. The adhesion between LDPE and Al is based mainly on acid-base interaction between oxidized groups on the LDPE surface and aluminium oxide and on van der Waals interactions between the two layers.[2-4] A carboxylic acid, sorbed by the LDPE, can interact with these groups and cause a decrease in interlayer adhesion or possibly a delamination. Most likely, water permeated through the LDPE together with the carboxylic acid and interfered with the adhesive forces between the two layers.

The recovery of adhesion in acetic acid seemed to be related to the formation of a layer of aluminium acetate and aluminium hydroxide. ATR analysis performed with FTIR on the inside of the LDPE film exposed to 3% acetic acid showed that the carboxylate concentration ($1587cm^{-1}$) and hydroxide concentration (1050 cm^{-1} and 3400 cm^{-1}) increased steadily during the storage time, indicating the presence of acetate and probably water. FTIR (RAS) and ESCA measurements on the Al foil showed a carboxylate and a hydroxide formation already after 8 days storage. Thus, the mechanism behind the recovery in adhesion was suggested to be as follows: Acetic acid and water permeated across the polymer producing H_3O^+. At the low pH (~ 2.5) the protective aluminium oxide layer was dissolved under reaction with acetic acid and water, forming aluminium acetate, aluminium hydroxide and hydrogen gas. The acid-base interactions between the oxidized aluminium and the oxidized LDPE were partly restored through acid-base interactions.[5,6] After a few weeks of storage, white spots were visible on the Al foil. They were believed to consist of Al acetate and Al hydroxide. Furthermore, the envelopes containing 3% acetic acid started to expand through gas formation after prolonged storage (3-4 months). Head-space GC analysis showed that the gas in the envelopes contained 88% H_2, 2% O_2 and 10% N_2.

3.1.2 Propionic Acid. Propionic acid did not delaminate LDPE and Al foil as rapidly as did acetic acid (Figure 1). A total delamination was observed on day 7, and did not recover during the rest of the observation period. FTIR (ATR) showed that the carboxylate and the hydroxide concentration in the LDPE increased. FTIR (RAS) analysis showed that the salt formation was slower on the Al surface exposed to propionic acid than on the Al exposed to acetic acid. In contrast to acetic acid, no gas formation was found for the material exposed to propionic acid.

A delamination induced by sorption occurred for propionic acid in a similar manner as for acetic acid. The level of corrosion of the Al surface, however, seemed to differ for the two acids. A carboxylate was visible already after 7 days of storage on the Al foil exposed to acetic acid, but on the Al foil exposed to propionic acid a carboxylate layer was not seen until day 28. Acetic acid was more corrosive than propionic acid, but, although a layer of propionate seemed to be formed in the latter case, the reaction with the aluminium foil was much less severe. It is believed that propionic acid formed a dense oxide layer on the Al surface, which retarded further production of propionate considerably, resulting in a weaker adhesive layer between the polymer and the aluminium.

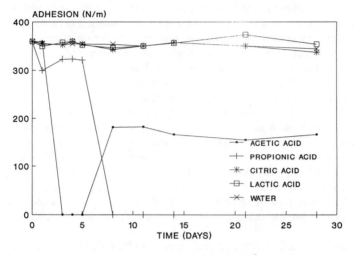

Figure 1 *Results of peel tests of laminates in Study II* [6]

3.1.3 Citric Acid and Lactic Acid. Citric acid, lactic acid and water did not affect adhesion significantly during the storage time (Figure 1). This was confirmed by FTIR (ATR) measurements on the LDPE, where no increase in the concentration of either carboxylate or hydroxide was observed.

The reason for this was probably the polarity and chemical structure of the acids, which make them much less soluble in LDPE than the aliphatic acids. Lactic acid carries an extra hydroxylic group, and citric acid is a bulky, highly polar molecule with three carboxylic groups and a hydroxylic group.

3.2 Fatty Acids

3.2.1 Fatty Acids of Various Chain Length. The effect of the saturated fatty acids: $C_{10:0}$, $C_{12:0}$, $C_{14:0}$, $C_{16:0}$ and $C_{18:0}$, on the interlayer adhesion between LDPE and Al foil were examined and compared with the effect of $C_{18:1}$ and $C_{18:2}$ (Study III).

The difference between the acids was small, although $C_{18:1}$ seemed to decrease the adhesion slightly more than the other acids (Figure 2). The interlayer adhesion decreased by ~ 40 % during the first week of storage and remained at that level during the rest of the storage time without causing delamination.

The sorption of fatty acids into the LDPE film reached a steady state in 1-2 days. The amount of sorbed fatty acid increased with increasing chain length and increasing saturation, as can be seen for $C_{18:0}$, which was sorbed in ~ 5 times the amount of $C_{10:0}$ and ~ 3 times the amount of $C_{18:1}$. The van der Waals interactions between the non-polar polymer and the aliphatic hydrocarbon chain increase with increasing chain length and therefore the solubility in the polymer increases. FTIR (ATR) measurements made on the

LDPE film indicated a considerable increase in carbonyl content for the material exposed to $C_{18:0}$. Contact angle measurements showed that the wettability of the Al foil decreased with increasing chain length and increasing degree of saturation. A possible explanation for the low sorption of the polyunsaturated fatty acids is that the extra double bonds make the molecule more rigid causing it to diffuse more slowly in the polymer matrix and it therefore builds up on or near the surface. This affects the concentration gradient, which is the driving force for the sorption process. The loss of adhesion did not seem to be proportional to the amount of sorbed fatty acid. Fatty acids $C_{14:0}$, $C_{16:0}$ and $C_{18:0}$ were sorbed in 3-5 times the amount of $C_{10:0}$, but the difference in the effect on adhesion was marginal. The major part of the absorbed fatty acid was most likely dissolved in the bulk of the polymer without interfering with the adhesive forces which are active only on the surface.

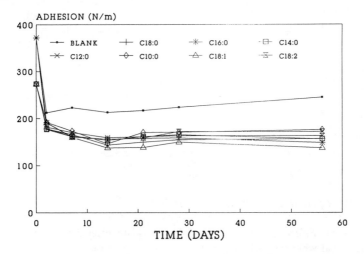

Figure 2 *Results for peel tests of laminates in Study III* [7]

3.2.2 Unsaturated Fatty Acids. In this study the effect of $C_{16:1}$, $C_{18:1}$, $C_{18:2}$ and $C_{18:3}$ on the interlayer adhesion was examined. The fatty acids were dispersed in a water/ethanol solution with SDS as an emulsifier.

The mono-unsaturated acids $C_{16:1}$ and $C_{18:1}$ caused a delamination between LDPE and Al in two days that did not recover during the rest of the experimental period (Figure 3). In the laminates exposed to $C_{18:2}$ and $C_{18:3}$ the adhesion decreased abruptly, but recovered somewhat and reached about half of its initial value after two weeks of storage. The adhesion was not uniform in the latter case, but varied from a total lack of adhesion to almost the same adhesion as in the untreated material in the same sample.

The sorption into the LDPE film reached a steady state the first 1-2 days. As was reported in the previous study, a considerable difference in sorption was observed for the mono-unsaturated acids and the polyunsaturated acids: 2-3 times the amounts of $C_{16:1}$ and $C_{18:1}$ were sorbed compared with $C_{18:2}$ and $C_{18:3}$. FTIR measurements showed a high concentration of carboxylate on the LDPE exposed to $C_{16:1}$ and $C_{18:1}$ and contact angle measurements also showed a lowered wettability of the Al foil. Contact angle measurements on the Al foils exposed to $C_{18:2}$ and $C_{18:3}$, on the other hand, showed increased wettability and FTIR showed a low carboxylate concentration in the LDPE.

It is believed that the free fatty acids and the SDS molecules form mixed micelles in equilibrium with a small amount of fatty acid dissolved in the water/ethanol phase. The mixture was quite clear and was considered to be a single phase system, or a solution, rather than an emulsion. The solution in Study IV, on the other hand, was a true solution, in which the fatty acid molecules were molecularly dissolved and evenly distributed in the solvent. As previously mentioned, the sorption of the fatty acids affects the adhesion by interfering with the active groups in the LDPE-Al interface. As was demonstrated in Study III, however, the fatty acid sorption alone does not explain the delamination. It is believed therefore that water permeates together with the unsaturated acids, resulting in a weak boundary layer, which leads to delamination.

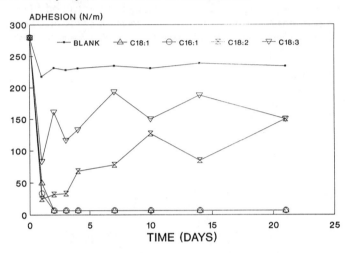

Figure 3 *Results of peel tests of laminates in Study IV* [7]

3.2.3 Synergistic Effects. In Study V possible synergistic effects between $C_{18:1}$, (oleic acid) and acetic acid were investigated.

The laminate exposed to $C_{18:1}$ and acetic acid was delaminated on day 4, while the laminate exposed to $C_{18:1}$ was delaminated first on day 7 (Figure 4).

In the emulsion with $C_{18:1}$, and the emulsion containing $C_{18:1}$ and acetic acid, a high rate of sorption occurred during the first 3-4 days, levelling off to a lower rate and finally approaching an equilibrium around day 21. The amount of $C_{18:1}$ sorbed by the LDPE almost doubled in the presence of acetic acid. In the emulsion containing $C_{18:1}$ and acetic acid, a formation of oil droplets was observed on day 4, indicating an instability in the emulsion.

A sorption-induced delamination occurred as previously described. The increased sorption of $C_{18:1}$ in the presence of acetic acid may be due to the increased contact between the oil and the LDPE film when the oil-droplets appeared. However, the increase started already on day 2, before the oil-droplets were seen, which indicates that there may be some sort of synergistic effect between $C_{18:1}$ and acetic acid.

The adhesion did not recover for the laminate exposed to $C_{18:1}$ and acetic acid. Contact angle measurements showed increased wettability of the surface of the Al foil exposed to acetic acid, due to the formation of an acetate. This was also seen with FTIR on the LDPE film and on the Al foil. On the Al foil exposed to $C_{18:1}$ and acetic acid a carboxylate was detected on the LDPE film and on the Al foil. However, contact angle measurements indicated an extremely hydrophobic layer on the Al foil. This means that, in spite of the formation of a carboxylate, the wettability decreased because of the penetrating fatty acid. This may explain why no recovery in adhesion was observed.

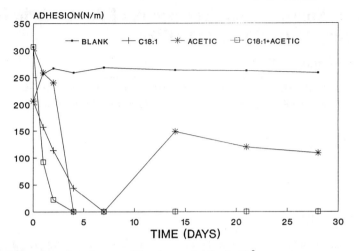

Figure 4 *Results of peel tests of laminates in Study V* [8]

References

1. M.A. Schroeder, B.R. Harte, J.R. Giacin and R.J. Hernandes, *J. Plastic Film & Sheeting,* 1990, **6**, 232.
2. F.M. Fowkes, 'Proceedings of the 34th international meeting of the Société de Chimie Physique, Paris', Ed. J.M. Georges , Elsevier, Oxford, 1982, p. 119.
3. K.W. Allen, *J. Adhesion,* 1987, **21**, 261.
4. T. Hjertberg, B.-Å. Sultan and E.M. Sörvik, *J. Appl. Polym. Sci.*, 1989, **37**, 1183.
5. G. Olafsson, M. Jägerstad, R. Öste, B. Wesslen, *J. Food Sci.,* 1993, **58**, 215.
6. G. Olafsson, M. Jägerstad, R. Öste, B. Wesslen and T. Hjertberg, *Food Chem.*, 1993, **47**, 227.
7. G. Olafsson and I. Hildingsson, submitted for publication, 1994.
8. G. Olafsson, I. Hildingsson, and B. Bergenståhl, submitted for publication, 1994.

Synergy in Sorption and Transport of Acetic Acid in Low Density Polyethylene (LDPE)

I. Hildingsson and B. Törnell

DIVISION OF CHEMICAL ENGINEERING II, CHEMICAL CENTER, LUND INSTITUTE OF TECHNOLOGY, PO BOX 124, S-22100 LUND, SWEDEN

1 INTRODUCTION

Large quantities of polymeric materials are used in food packaging. As polymeric materials are not inert and may interact with the packed food, it is obvious that interactions between food components and polymeric packaging materials must be considered in the development of safe packaging systems.

Much of the research on interactions between packed food and their packages has dealt with effects exerted by the package on the food, such as migration of various substances from the packaging material into the food, and aroma changes due to sorption of flavour substances into the packaging material. Food components however, may also affect the properties of packaging materials and the package as a construction. In certain cases, interactions of this kind may endanger the integrity of the package. It has been reported that migration of limonene into polyethylene affects the mechanical properties of the polymer.[1] It has also been reported that certain food products may cause adhesion failures in laminates containing an internal Al-foil gas barrier.[2] Recently, it was shown that certain organic acids and fatty acids caused a rapid loss of adhesion at the Al-PE interface in a laminate of a type commonly used for safe packaging of milk products.[3-5] There are also indications that a certain food component may be more aggressive in a particular food product than in others, pointing to the importance of synergistic interactions between individual food components in the product.

Presently, we are engaged in model studies set up to elucidate mechanisms which may explain the role of synergistic interactions between food components in food-package interactions. In this study, we report a comparison between the transport rate of acetic acid through LDPE films from liquid food simulants, either free from or containing hydrogen bond acceptor substances. The hydrogen bond acceptor substances used were butanol, valeraldehyde, 3-pentanone and oleic acid. The effect of hydrogen bond acceptor substances on the distribution of acetic acid between water and a nonpolar liquid, octane, is also reported. The liquid food simulants used were either pure aqueous solutions, or, in the case of oleic acid, oil in water emulsions.

2 MATERIALS AND METHODS

2.1 Materials

The chemicals used in the experiments were acetic acid (Analytical grade, Merck, octane, 99+% Janssen Chimica), oleic acid, (99+% Laurodan Chemicals), 3-pentanone, (99% BHD) and valeraldehyde (99% Aldrich). The water was of HPLC quality, and was degassed by boiling before use. Two different LDPE films, kindly supplied by Tetra Laval AB, were used in the measurements. These films were reportedly prepared from slightly different LDPE materials from Exxon. One of the films was a blown film with a thickness of 19μm, the other an extruded film of thickness 43 μm.

2.2 Permeation Studies

2.2.1 *Acetic acid.* Rates of permeation of acetic acid (HAc) through LDPE films were measured at room temperature ($21 \pm 0.5°C$), using liquid permeation cells consisting of two continuously stirred glass chambers, each with a capacity of 55 ml, separated by the polymer film under study. In all experiments, except for the content of acetic acid, the two cell chambers had the same starting compositions. The rate of acetic acid permeation was evaluated by following the increase in concentration of acetic acid in the receiving chamber with time. The determination of acetic acid was facilitated by using ^{14}C-tagged acetic acid (Sigma Chemical Co) and liquid scintillation counting. In these determinations, a 450 μl sample of the receiving chamber solution was mixed with 5 ml scintillation liquid (Ultima gold, Packard) in a 6 ml plastic vial, and the radioactivity counted in a Rackbeta 1219 (LKB) liquid scintillation counter for 10 min.

In the experiments with butanol, valeraldehyde and 3-pentanone, the solvent was pure water and the polymer a blown 19 μm LDPE film.

In the experiments with oleic acid, the liquid phase was a 10 % rape-seed oil in water emulsion (see below), either free from, or containing 1 % oleic acid (10 % with respect to the oil phase). In these experiments an extruded LDPE-film with a thickness of 43 μm was used.

The permeation coefficient P, (cm^2/s) was calculated according to equation 1

$$P = \frac{\text{(quantity of permeant) (film thickness)}}{\text{(film area) (time) (concentration difference across the film)}} \quad [1]$$

The water in oil emulsions were prepared from refined rape-seed oil (Karlshamn Oil & Fat AB), containing less than 0.1 % free fatty acids, by first mixing 10 parts (by weight) of rape-seed oil, or 10 parts of a mixture of 90 parts of rape-seed oil and 10 parts of oleic acid, with 0.3 parts of an ethoxylated sorbitan oleate emulsifier (Tween 80, HLB 15, ICI Specialty Chemicals). After adding 90 parts of water to the oil phase, the mixture was preemulsified using an Ultra Turrax (IKA, Germany) high speed stirrer. The final emulsification step was performed by passing the emulsion two times through a high pressure homogenizer (Cook & Laguce, Microfluidizer[TM] Inc., Newton, MA, USA), using a pressure difference of 1000 bar and room temperature. After emulsification, acetic acid was added to the emulsion samples to be used as leaving cell chamber solution.

2.2.2 *Valeraldehyde and 3-pentanone.* The rate of permeation of 3-pentanone and valeraldehyde through LDPE film from pure water solutions was measured using the same liquid permeation cell as described above, and the blown 0.19 μm LDPE film. The rate of permeation was evaluated by following the increase in concentration of 3-pentanone/valeraldehyde in the receiving cell solution with time. The concentration of 3-pentanone/valeraldehyde was determined by gas chromatography (Perkin Elmer F 33) using a packed column (OV-225). Ethanol was used as internal standard and the injection volume was 4.0 μl.

2.3 Distribution Coefficients

The distribution coefficient of acetic acid between water and octane was measured using pure acetic acid solutions or equimolar mixtures of acetic acid and 3-pentanone. In these experiments, 4 ml water was mixed with 4 ml octane in glass vials. Measured amounts of acetic acid, or a 1:1 (on a mole basis) mixtures of acetic acid and 3-pentanone were added. The vials were shaken until equilibrium was reached and subsequently centrifuged for 20 minutes at 3000 rpm. The concentration of acetic acid in the octane phase was measured by gas chromatography (Varian 3400) using a Nukol wide bore capillary column from Supelco. The amount of 3-pentanone in the octane and water phases was measured by gas chromatography (Perkin Elmer F 33) using a packed column (OV-225). Ethanol was used as internal standard and the injection volume was 4.0 μl.

3 RESULTS AND DISCUSSION

The permeation coefficient for acetic acid through polyethylene was found to increase with an increase in the acetic acid concentration (see Figure 1 and Table 1), indicating that the formation of acetic acid dimers in the polyethylene phase is an important factor in explaining the interaction between polyethylene and aqueous solutions of acetic acid. This conclusion is supported by the observation that the distribution coefficient of acetic acid between pure water and octane increases with an increase in acetic acid concentration (Figure 2). The formation of acetic acid dimers in organic solvents is well established.[6]

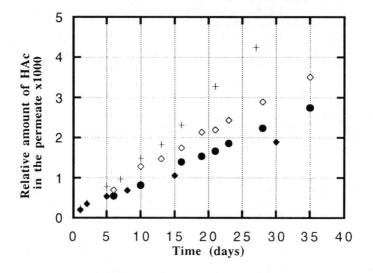

Figure 1. *Permeation of acetic acid from aqueous solutions containing ♦ 0.1, ● 0.2, ◊ 0.4, and + 0.5 mol/l of acetic acid, through a blown film of LDPE with a thickness of 19 μm, at 21°C.*

Table 1. *Influence of Acetic Acid Concentration and the Presence of 0.05 M 3-Pentanone or 0.05 M Valeraldehyde on the Permeation Coefficient, P, for Transport of Acetic Acid from Aqueous Solutions through Blown Films of LDPE at Room Temperature.*

[HAc]	$P \times 10^{12}$ / $(cm^2/s$		
M	*no additives*	*3-pentanone*	*valeraldehyde*
0.1 0.2 0.4 0.5	4.07 5.37 6.68 10.87	6.49	10.80

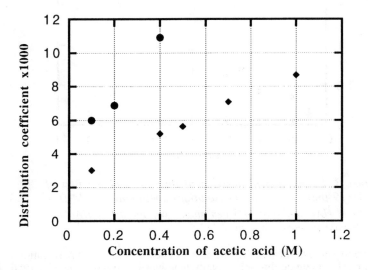

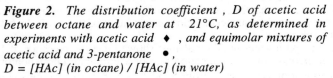

Figure 2. *The distribution coefficient , D of acetic acid between octane and water at 21°C, as determined in experiments with acetic acid ◆ , and equimolar mixtures of acetic acid and 3-pentanone ● ,*
$D = [HAc]$ *(in octane) /* $[HAc]$ *(in water)*

The effect of the presence of hydrogen bond acceptor substances on the permeation rate of acetic acid through polyethylene is exemplified by the results in Figure 3 and 4. It is evident that the presence of valeraldehyde and 3-pentanone caused the permeation rate of acetic acid to increase (Figure 3). Similar effects have been found in experiments with butanol and methyl ethyl ketone as hydrogen bond acceptor substances. The increase in the distribution coefficient of acetic acid between water and octane produced by adding 3-pentanone to the system (Figure 2), strongly suggests that the hydrogen bond acceptors affected the permeation rate of acetic acid through polyethylene by forming hydrogen bond complexes in the polymer phase. By this type of complex formation, the effective polarity of acetic acid decreases, leading to an increase in the total acetic acid concentration in the polymer phase. The fact that the increased solubility is paralleled by an increase in permeation rate, suggests that the increased solubility of acetic acid in polyethylene more than compensates for the possible effect of complex formation on lowering the effective diffusion coefficient of acetic acid.

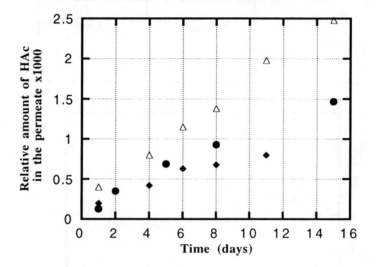

Figure 3. *Permeation of acetic acid through a blown film of LDPE (19 µm) , from aqueous solutions containing 0.1 M HAc* ♦ *, 0.1 M HAc and 0.05 M 3-pentanone* • *, and 0.1 M HAc and 0.05 M valeraldehyde* ∆ *, at 21°C.*

Separate experiments showed that the effect of acetic acid on the permeation rate of 3-pentanone or valeraldehyde through polyethylene was too small to be observed. This can be explained by the fact that the permeation coefficient of these substances were found to exceed that of acetic acid (at an acetic acid concentration of 0.1 mol/l) by a factor of 2,300 and 5,600 respectively.

The experiments with water in oil emulsions showed that the presence of 1% oleic acid produced an increase in the permeation rate of acetic acid, corresponding to an increase in the permeation coefficient from 5.7×10^{-12} to 8.2×10^{-12} cm^2/s (Figure 4). Other experiments, not discussed here, showed that oleic acid was fairly rapidly sorbed into the polyethylene film from the same oil in water emulsion. In this case also, it is likely, that the higher rate of permeation of acetic acid can be explained by to an increase in the total acetic acid concentration in the polymer.

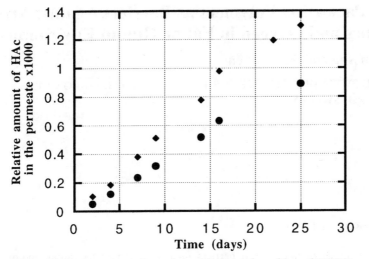

Figure 4. *Permeation of acetic acid through a film of LDPE (43 μm), from rape-seed oil in water emulsions containing 2.7 % (w/w) HAc ● , and 2.7 % HAc and 1% oleic acid ◆ , at 21°C.*

4 CONCLUSIONS

The present study has shown that acetic acid permeates more rapidly through polyethylene from aqueous acetic acid solutions if these solutions also contain a hydrogen bond acceptor, which may form acid-base complexes with acetic acid in a hydrophobic environment.

Acknowledgements

This project was financially supported by the Swedish Council for Forestry and Agricultural Research (SJFR), The Swedish National Board for Technical and Industrial development (NUTEK) and the following companies Tetra Laval AB, Neste polyethylene, PLM, STORA, AB Felix, Swedish Nestlé and the Swedish Meat Research Institute. The authors are indebted to Dr M. Jägerstad and Techn. lic. G. Olafsson, Chemical Centre, Div. of Food Chemistry, and to Mikael Berlin and Kerstin Petersén, Tetra Laval AB, for valuable discussions.

References

1. B.R. Harte, J.R. Giacin, T. Imai, J.B. Konczal, & H: Hoojjat. Food Packaging Technology ASTM STP 1113, D Henyon, Ed., American Society for Testing and Materials, Philadelphia, 1991, 18.
2. M.A. Schroeder, B.R. Harte, J.R. Giacin & R.J. Hernandes, *J. Plastic Film & Sheeting*, 1990, **6**, 232.
3. G. Olafsson, M.Jägerstad, R. Öste, B.Wesslen & T. Hjertberg, *Food Chem.*, 1993, **47**, 227.
4. G. Olafsson, M. Jägerstad, R. Öste & B.Wesslen, *J. Food Sci.*, 1993, **58**, 215.
5. G. Olafsson, I. Hildingsson, 1994. To be published.
6. F. C. Rubio, V.B. Rodriguez & E. J. Alamenda, *Ind. Eng. Chem. Fundam.*, 1986, **25**, 142.

Food Packaging Polymers as Barriers Against Aroma Vapours and Oxygen in Fat or Humid Environments

F. Johansson and A. Leufvén

SIK, THE SWEDISH INSTITUTE FOR FOOD RESEARCH, PO BOX 5401, S-40229 GÖTEBORG, SWEDEN

1. INTRODUCTION

Polymer materials are not absolute barriers against oxygen, water vapour and organic materials, but are widely used in food packaging applications. This has led to a demand for better understanding of the factors that influence the transport mechanism through polymer materials. A number of factors influence the barrier properties of a packaging material. These factors are to be found both in the food itself and in the environment. The type of polymer and aroma compound affect the degree of sorption. Furthermore, the composition of the packed food (e.g. fat content, pH, pulp content, and types of aroma compounds present) may have an influence on the sorption characteristics of the packaging material. Environmental factors, such as the temperature and relative humidity, may affect the barrier characteristics of the packaging material.

A substantial oxygen barrier is a prerequisite in many food packaging applications, as even a small amount of oxygen may have an adverse effect on the sensory quality of the product. Determinations of water vapour and oxygen transmission rates have been made for a wide range of packaging materials[1]. However, contact with foods may alter the barrier characteristics of a packaging material. It is thus important to study the barrier performance of the material under realistic conditions. These are particularly difficult to obtain if the barrier characteristics of a packaging material vary over the expected shelf-life of the packed product.

This contribution will discuss two aspects that may influence the barrier characteristics of a polymer packaging material; firstly, the effect of relative humidity (RH) on the sorption and permeation behaviour of aroma compounds (i.e. aldehydes and alcohols) in some polymers used for food packaging[2]; secondly, the effect of a vegetable oil (i.e. rapeseed oil) on the oxygen barrier properties of some different polymer packaging materials[3]. Furthermore, the combined effect of sorbed oil and different relative humidities on the oxygen transmission rate (OTR) through the packaging materials was investigated in order to study the synergistic or antagonistic effects of two possible causes of altered barrier performance.

2. METHODS

The permeation measurements were made with aroma compounds in their vapour state using a permeation cell[4] in which the investigated polymer film was placed. The permeation of aroma compounds through the film was followed by collecting the permeated compounds on an adsorbent material (Chromosorb 102) with subsequent thermal desorption of the compounds on a gas chromatography column. The aroma compounds sorbed in the polymer films were extracted by supercritical carbon dioxide[5] and quantified by a gas chromatograph equipped with a flame ionisation detector (GC-FID). The polymer films used were HDPE, high-density polyethylene, 40 μm; LLDPE, linear low-density polyethylene, 40 μm; and EVOH, ethylene vinyl alcohol copolymer, 12 μm. The aroma compounds used were straight-chain aliphatic aldehydes C_4 - C_8, and C_{10}, and primary and secondary alcohols (C_5 - C_7).

The oxygen transmission rate was determined according to ASTM D3985-81 using a MOCON Oxtran 2/20, at four relative humidities; 0, 35, 55, and 95%RH. All tests were performed at 25°C. The polymer films used were HDPE, high density polyethylene, 40 μm; PP, polypropylene 300 μm, and APET, amorphous polyethylene terephthalate, 300 μm. All polymer films were tested both without and with storage in rapeseed oil for 20 and 40 days. The amount of sorbed oil in the films was determined by extracting the films using Soxhlet extraction equipment.

3. THE EFFECT OF THE RELATIVE HUMIDITY ON THE PERMEABILITY OF AROMA VAPOURS THROUGH POLYMER FILMS

Different classes of aroma compounds exhibit different permeation and sorption behaviour at different relative humidities. In the polar EVOH film, which is known to be strongly affected by the RH, the permeability of aldehydes increased with increasing RH, Figure 1. The permeability of alcohols, on the other hand, decreased at high RH (i.e. 95%RH), while the sorption of the alcohols increased with increasing RH. The sorption of aldehydes was slightly decreased at 90% RH. This is explained by the combination of the plasticising effect of the water and the interaction between the aroma and water vapours. The increased permeability of the aldehydes may be explained by the swelling of the polymer caused by the sorption of water. The aldehydes do not have the same ability to interact with the water as the alcohols.

The sorption and permeation of aroma compounds in the non-polar polyethylene films studied were also affected, but to a smaller extent.

4. THE EFFECT OF SORPTION OF A VEGETABLE OIL INTO POLYMER FILMS ON THE OXYGEN TRANSMISSION RATE

The OTR of the HDPE film was significantly increased, compared with the virginal polymer film, after being stored for 20 days in rapeseed oil. The OTR of the HDPE film was significantly influenced by the time that the polymer had been in contact with the oil (i.e. 20 or 40 days); Figure 2. The OTR of the HDPE film soaked for 40 days in the oil was increased, compared to the virginal polymer, by between 37 and 45%, depending on the RH. There was approximately 2.3 times more rapeseed oil sorbed in the HDPE films stored for 40 days than in the HDPE films stored for 20 days.

There was no change in the OTR of the PP film after 20 days of storage. After 40 days of storage in rapeseed oil, however, the OTR was significantly increased compared to the virginal film. The increase in OTR was greatest at dry conditions and became smaller with increasing RH. At 95%RH, no significant difference was seen between the virginal polymer and the polymer containing sorbed oil.

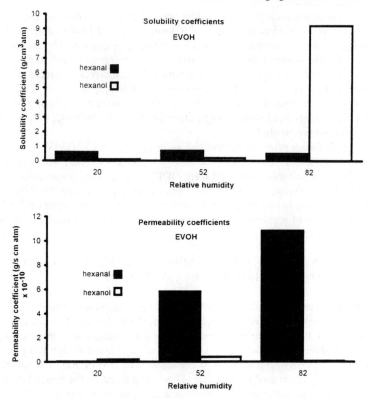

Figure 1. *The solubility and permeability coefficients of hexanal and hexanol in EVOH at different relative humidities.*

The OTR of the APET film was slightly decreased with increased RH, both in the virginal polymer and in the polymer containing sorbed rapeseed oil. The decrease in OTR was, however, counteracted to some extent by the swelling effect of the sorbed oil at higher RH.

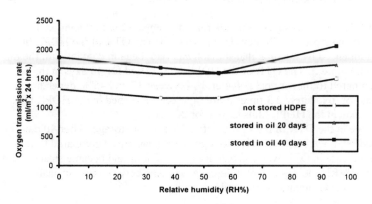

Figure 2. *The OTR of HDPE with and without sorbed rapeseed oil.*

5. CONCLUSIONS

In some cases, the RH has a considerable influence on the permeation and sorption characteristics of aroma vapours in the investigated films (i.e. LLDPE, HDPE, and EVOH). This factor should thus be taken into consideration when choosing a package material for a given application. The differences in sorption and permeation behaviour between the aroma compounds investigated (i.e aldehydes and alcohols) are interpreted as the result of the combination of a plasticising effect of water and the interactions between aroma and water vapours.

The APET film remained an excellent oxygen barrier even after storage in rapseed oil for 40 days. The polyolefins (i.e. HDPE and PP) showed an increased OTR after being stored in rapeseed oil. This was attributed to the swelling effect of the sorbed oil.

References

1. R.W. Tock, *Advances in Polym. Technol.*, 1983, **3**, 223.
2. F. Johansson and A. Leufvén, *J. Food Sci.*, Submitted.
3. F. Johansson and A. Leufvén, *Pack. Technol. Sci.*, Submitted.
4. A. Leufvén and U. Stöllman, *Z. Lebensm. Unters. Forsch.*, **194**, 355.
5. F. Johansson, A. Leufvén and M. Eskilson, *J. Sci. Food Agric.*, **61**, 241.

Extractable Components in Paperboard

M. Björklund Jansson, A.-K. Bergqvist, and N.-O. Nilvebrant

STFI, SWEDISH PULP AND PAPER RESEARCH INSTITUTE, BOX 5604, S-11486
STOCKHOLM, SWEDEN

1 ABSTRACT

Supercritical carbon dioxide has been used for extraction of components, mainly wood
extractives and wet strength resin, e.g. rosin size and alkyl ketene dimers, from paper. A
method to study the redistribution of wood extractives between different layers in paperboard and/or between paperboard and a plastic film has been developed. The importance
of variables such as time, temperature, vapour pressure and adhesion to the fiber phase
and the plastic layer have been evaluated for different types of extractives.

2 INTRODUCTION

Extractable components in pulp normally consist of residues of wood extractives. In
paper, depending on the paper quality, extractable substances may also have been added
in the paper production process in order to achieve a specified quality. In laminated board
extractives may also be derived from the plastic layer. The amounts of extractable compounds are defined by the choice of solvent and the procedure for extraction. In this summary only lipophilic extractives extractable with an organic solvent or supercritical carbon dioxide will be discussed.

Lipophilic wood extractives have mainly been studied because they cause problems
in the pulp and paper production. They are difficult to remove in the pulp washing stages
and they are known to give sticky deposits on process equipment. Time dependent redistribution of lipophilic extractives, so called self sizing, is also known to effect the surface
properties, especially the water resistance and friction, of uncoated paper.

The redistribution of extractives in packages e.g. between different layers in multilayer board material must also be considered with respect to the concept of functional barrier. In order to understand these processes fully more basic knowledge of the transport of
extractives between layers with different character is needed.

3 EXTRACTABLE COMPONENTS IN PAPER

3.1 Wood extractives

Main classes of lipophilic extractives in wood include fatty and resin acids, sterols,
fatty alcohols and esters, e.g. triglycerides and sterylesters. Further alkanes and several
terpenoid compounds of different size, from monoterpenes to polyprenoids, have been
identified. The total amount of extractives as well as the extract composition varies significantly for different wood species.

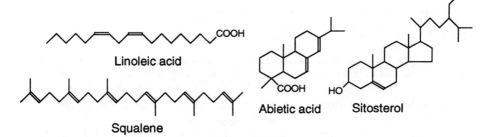

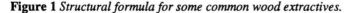

Figure 1 *Structural formula for some common wood extractives.*

In the kraft pulping process the esters will be hydrolysed and most of the extractives will be removed in the several washing stages. A bleached softwood kraft pulp normally has an acetone extract content around or below 0.1 %. The corresponding value for a birch kraft pulp is somewhat higher, around 0.2 to 0.3 %. As distinguished from kraft pulps extracts from mechanical pulps will contain also unhydrolysed esters. Examples of some common wood extractives are shown in Figure 1.

During the pulp bleaching stages the bleaching agents will to some extent react with the extractives. For older type of bleached pulp, bleached with chlorine gas, chlorinated wood extractives have been documented. In the pulp bleaching processes used today in Sweden chlorine has been replaced by chlorine dioxide or totally chlorine free bleaching agents such as e.g. hydrogen peroxide or ozone. All these will primarily react with wood extractives by oxidation and chlorinated compounds are not formed.

3.2 Paper additives

In the paper mill chemicals are used either to enhance the paper quality or to improve the production efficiency. Examples of the first type are wet strength resins, paper sizes (=hydrophobation chemicals) and fillers. Examples of the second type are defoamers, retention aids and tensides used for controlling deposits on the machinery. In order to improve the printability of the paper different types of coating, e.g. clay coating, are often applied on the paper surface.

Traditionally a combination of resin acids, called rosin size, and aluminium sulphate have been used to make the fibers more hydrophobic. Today many paper mills instead use synthetic sizing agents, such as alkyl ketene dimers. Both types of sizing agents are readily extractable from paper or paperboard with organic solvents or supercritical carbon dioxide. Below a method based on supercritical chromatography for analysis of paper AKD or rosin size is described.

3.3 Extractives in plastic films

Paper qualities laminated or coated with polymeric materials also contain extractable components from the polymer such as monomers, and additives used in the plastic production. These will not be discussed further. One example of analysis of extracts from polypropylene with supercritical chromatography is given by Bücherl et al.[1]

4 EXTRACTION WITH SUPERCRITICAL CARBON DIOXIDE

Above the critical point the density of gas and liquid is the same and a substance in this state is defined as a supercritical fluid. Supercritical fluids have solute diffusivities an order of magnitude higher and viscosities an order of magnitude lower than liquid solvents, which results in a very effective mass transfer. In combination with a density and a solvating capacity comparable to liquids this gives supercritical fluids their special extraction potential.

Supercritical carbon dioxide is the most commonly used solvent for SFE partly because it has a relatively low critical pressure (7.4 MPa) and a critical temperature of only 31°C. In addition, carbon dioxide is nontoxic, nonflammable, chemically inert and comparatively inexpensive.

The established method for determination of extractives in pulp or paper is based on Soxhlet extraction with an organic solvent. An interesting alternative to Soxhlet extraction is extraction with supercritical carbon dioxide (SF-CO₂). A method for extraction of

pulp has earlier been described.[2] At optimal conditions SF-CO₂ gave a higher yield of typical wood extractives compared to Soxhlet extraction with acetone or dichloromethane.

The same extraction conditions as used for pulp have also been used for extraction of paper samples.

5 SUPERCRITICAL FLUID CHROMATOGRAPHY (SFC)

Supercritical carbon dioxide is also used in supercritical fluid chromatography (SFC). This technique is today well known as a complement to GC and HPLC and it is well suited for the separation of complex mixtures of oleophilic compounds, such as wood extractives. The main advantage compared to GC is the possibility to separate compounds with higher boiling points or thermally labile compounds. As in GC a flame ionisation detector (FID) can be used. SFC can be used alone or in combination with on-line or off-line SFE.

The conventional analytical technique for wood resin, based on GC, gives a separation of resin components in individual compounds. However, for many applications there is no need for such detailed information, whereas on the other hand a quantitative method giving information of the different classes of compounds present in the extract would be of significant value. The need for hydrolysis of esters and derivatisation of acids also makes GC-analyses less attractive.

An analytical method based on supercritical fluid chromatography for group separation of wood resin components have been developed at STFI.[2] With a rapid density programme a separation of compounds representing different classes of wood resin components can be achieved as illustrated for some model compounds in Figure 2.

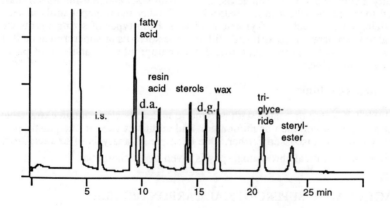

Figure 2 *Supercritical fluid chromatogram of model compounds.*
D.a. = diterpene alcohol, d.g. = diglyceride , i.s. = internal standard.

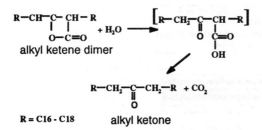

Scheme 1 *Hydrolysis of AKD.*

The method described above for analysis of wood extractives in pulp and paper samples can also be used for determination of alkyl ketene dimers (AKD) and rosin size in paper. AKD gives a typical pattern of normally three peaks, representing AKDs with different length of the alkyl chains.

During the drying of the sized paper some AKD will be reacted with the hydroxyl groups in cellulose. However some AKD will also be hydrolysed giving the corresponding alkyl ketones as shown in scheme 1. In the developed analytical method the ketones are separated from the AKD.

Figure 3 shows SFC chromatograms from SF-CO$_2$ extract from two different paper samples. In the chromatogram showing the SF-CO$_2$ extract of the AKD-sized paper sample only three peaks corresponding to alkyl ketones, (R=16+16, R=16+18, and R=18+18) can be seen.

6 REDISTRIBUTION OF WOOD EXTRACTIVES

A method to study factors influencing the redistribution of extractives between different layers in a multilayer paperboard has been developed. Factors studied include time, temperature, vapour pressure of the extractives and the affinity to the solid phase (fibers or plastic).

Several layers of preextracted paperboard are put together in a bundle, containing also a board prepared with the addition of the extractives to be studied. Also plastic films can be used in the bundles, as illustratedinFigure 4. The bundle is pressed together, wrapped in aluminium foil and stored at a selected temperature. After a selected period of time the bundle is broken and the different layers are extracted separately and analysed for the added extractives.

In one experiment pentadecane and two fatty acids, lauric acid and stearic acid, were added to the middle paperboard as model compounds for pulp extractives with different polarity and vapour pressure. The relative percentages of these compounds found left in

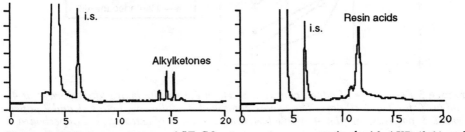

Figure 3 *SFC-chromatogram of SF-CO$_2$ extract from paper sized with AKD (left) and rosin size (right) respectively.*

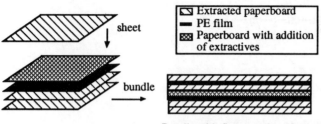

Figure 4. *Paperboard sheets and polyethene films are put together in a bundle.*

the middle paperboard after storage at 40 and 60 °C for 1 respectively 10 days are shown in Figure 5 for a bundle containing also two polyethene films.

The presence of polyethene films in a bundle markedly influences the relative distribution of the three compounds between the layers, and especially the distribution of pentadecane. In bundles containing polyethene films the amount of pentadecane found left in the middle paperboard was very small. Most of the added pentadecane was instead found in the polyethene films, as illustrated in Figure 6, showing the relative distribution of pentadecane and lauric acid between the different layers after storage at 60°C for 1 day. Figure 7 shows as a comparison the corresponding distribution of pentadecane and lauric acid in a bundle without polyethene films after the same storage conditions.

The results indicate that the vapour pressure of the extractives as well as the temperature determines the time to reach equilibrium conditions. However, the affinity to the solid phase seems to be the most important factor for the equilibrium distribution.

7 SUMMARY

Extraction with supercritical carbon dioxide (SFE) followed by off line supercritical chromatography (SFC) is a rapid and informative method for determination of extractable components in pulp and paper samples.

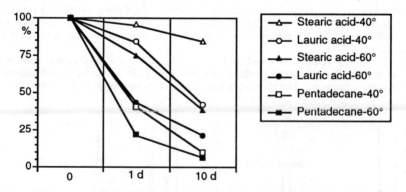

Figure 5 *Percentage of added compounds left in the middle board after storage at 40 and 60°C for 1 and 10 days. The bundle contained the following layers: 1 and 2 paperboard, 3 polyethylene film, 4 middle board with addition of extractives, 5 paperboard, 6 polyethene and 7 paperboard.*

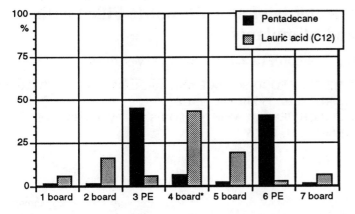

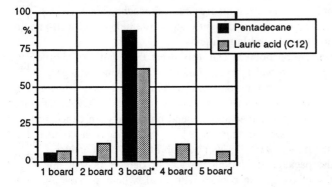

Figure 6 *Relative distribution of pentadecane and lauric acid in a bundle containing 5 layers of paperboard and two polyethene films, after storage at 60 °C for 1 day (24 hours).Before storage 100 % of the pentadecane and lauric acid were added to the middle board marked with *.*

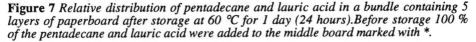

Figure 7 *Relative distribution of pentadecane and lauric acid in a bundle containing 5 layers of paperboard after storage at 60 °C for 1 day (24 hours).Before storage 100 % of the pentadecane and lauric acid were added to the middle board marked with *.*

A method for analysis of paper sizing agents, such as alkyl ketene dimers and rosin size based on SFE and SFC has been developed.

A method to study factors affecting the redistribution of extractives between different layers in a multilayer packaging material has been developed. The affinity to the solid phase seems to be the most important factor for controlling the time and temperature dependent redistribution between the layers.

References

1. Bücherl, T., Baner, A.L. and Piringer, O.-G.
    Deutsche Lebensmittel-Rundschau, 1993 **89** 69

2. Björklund Jansson, M., Dahlman,O., Månsson, K., Rutquist, A. and Wickholm J.
    Proceedings International Pulp Bleaching Conf,
    Stockholm, June 1991, p 123

Oxygen Permeability of an *OPP* Bottle Filled with Different Food Stimulants

L. Piergiovanni[1], P. Nicoli[1], L. Tinelli[1], and G. Vestrucci[2]

[1]DISTAM, DEPARTMENT OF FOOD SCIENCE AND MICROBIOLOGY, UNIVERSITY OF MILAN, V. CELORIA, 2 20133 MILAN, ITALY
[2]CSI, APPLIED RESEARCH CENTRE, MONTEDISON S.P.A., V. LOMBARDIA, 20 20021 BOLLATE (MILAN), ITALY

1 INTRODUCTION

The sensorial, hygienic and nutritional quality of beverages and liquid foods is frequently affected by the reactions which can take place in the presence of oxygen. Lipid oxidation, the auto-catalytic free radical mechanism in which fats are subject to direct attack by oxygen, is certainly the most studied and well-known deterioration phenomenon for oils and fatty liquids [1,2].

Other oxygen consuming phenomena can also affect the quality of bottled or packed foods; for instance microbial growth (mineral water, wine, vinegar, milk, fruit juices, sauces...), vitamin denaturation (milk, juices...), pigment deterioration (beer, wine, vinegar, juices...), enzymatic browning (juices...), flavour tainting (all). The rate and the extent of the reactions depend on the oxygen concentration in the medium and/or in the surrounding space [3,4]. Packaging can control this important variable at least in two ways: by its permeability and its geometry.

Permeability of the package (including leakage or defects in seal integrity) modulates the availability of oxygen in contact with food, whereas shape and size of the package establish the head-space (usually an air volume) and the ratio between the oxygen permeable surface and the mass of the sensitive product.

It has been reported [5] that liquids in contact with plastic films can reduce gas permeation to different extents, according to the type of polymer employed and to the kind of liquid, resulting in a particular form of interaction between food and packaging materials. This fact should be taken into account in selecting the kind of plastic bottle to be used for an oxygen sensitive product.

In this work, the oxygen permeability of a one-litre oriented polypropylene bottle, filled with solutions simulating different media and exhibiting different ratios of permeable surface/volume of liquid, was measured in order to get information about performances of the container in contact with a sensitive product.

2 MATERIALS AND METHODS

The plastic bottles were experimental containers made of PP copolymer, produced by Injection Stretch Blow Moulding process (ISBM) in a CSI-Montedison laboratory; the capacity of the bottles was 1 l, the weight was 30 g and the permeable surface was 5.5 dm^2; the oxygen and carbon dioxide permeability of empty containers, as measured by isostatic method (MOCON Instruments, Minneapolis, USA), were 11.3 and 32 (cm^3 /day bar), respectively.

The bottles were filled with different solutions: distilled water (DW), acetic acid 10 and 20% v/v in water (AA1, AA2), ethanol 50 and 70% v/v in water (ET5, ET7), saccharose in water at 10, 20% and 40% m/v (SA1, SA2, SA4); all the simulants were added with an antibacterial agent (mercuric iodide) to prevent oxygen consumption by micro-organisms.

The solutions were completely deaerated by purging with Argon and then a Clark's electrode (WTW mod. OXI 96, Weilheim, Germany) was hermetically sealed, excluding any void volume, into the neck of the bottle; the probe resulted symmetrically 3 cm from the walls, 4.5 cm from the mouth and 22.5 cm from the bottom of the bottle; the containers were subsequently put in a barrier bag (GRACE Italiana BB4L, Rho, Italy) filled with 99.999% oxygen (see Figure 1).

Four bottles were filled both with distilled water and glass balls (diameter 1 cm) in order to increase the ratio of "permeable surface/product volume" (these samples are indicated as LV=lower volume) and other four bottles were tightly wrapped with aluminium foil to reduce by one third the permeable surface for the same volume of water (these samples are indicated as LS=lower surface). The dissolving gas was therefore continuously monitored, in static conditions and at constant temperature (22 ± 1 °C), by means of a recorder (LINSEIS G8405, Bolzano, Italy) connected to the electrode. The gas permeation values were calculated from the slope of the linear regression of the dissolved oxygen versus time and expressed as cm^3/bottle day bar [6].

Figure 1 *Device used for the oxygen permeability measurements.*

The oxygen tightness of the system was tested by measuring the oxygen permeated through a returnable high barrier PET bottle and found satisfactory.

The oxygen solubility in the simulants was measured at approximately the same temperature (22 $\pm$ 0.5 °C) and with the same electrode. An oxygen flow of 40 l/min was continuously stirred in 200 cm³ of each solution and the gas dissolved was recorded as mg/l bar when the equilibrium had been reached.

3 RESULTS AND DISCUSSION

The oxygen solubilities in the different simulants are reported in Table 1. Only a high concentration of sugar reduced the oxygen solubility of distilled water; evidently the addition of solids decreases the oxygen solubility much more than the organic liquids (acetic acid and ethanol) which showed values ranging between 34 and 36 mg/l bar, quite close to the solubility of distilled water.

Table 1	Oxygen Solubility in Different Media
Food Simulant	*Solubility* (mg/l bar at 22$\pm$0.5 °C)
Distilled water	**36.5 $\pm$ 0.2**
Ethanol 50% in water (v/v)	**35.3 $\pm$ 0.4**
Ethanol 70% in water (v/v)	**36.5 $\pm$ 0.2**
Acetic Acid 10% in water (m/v)	**36.0 $\pm$ 0.2**
Acetic Acid 20% in water (m/v)	**36.2 $\pm$ 0.2**
Saccharose 10% in water (m/v)	**35.7 $\pm$ 0.1**
Saccharose 20% in water (m/v)	**34.2 $\pm$ 0.3**
Saccharose 40% in water (m/v)	**30.0 $\pm$ 0.2**

The results of progressive oxygen dissolution in distilled water and in 40% saccharose solution, measured in three different bottles, are reported in Figures 2 and 3. All the permeability data are summarised in Table 2 with the *determination coefficient* (R^2) of the linear regression calculated.

Table 2 Oxygen Permeability of Empty and Filled OPP Bottles with Different Food Simulants		
Sample	*Oxygen Permeability* (cm³ /bottle day bar at 23 $\pm$ 1°C)	R^2
Empty bottle	**11.30**	
DW (filled with distilled water)	**2.15**	**0.932**
AA1 (filled with Ac. Acid 10%)	**2.15**	**0.923**
AA2 (filled with Ac. Acid 20%)	**2.20**	**0.973**
ET5 (filled with ethanol 50%)	**2.16**	**0.979**
ET7 (filled with ethanol 70%)	**2.38**	**0.970**
SA1 (filled with saccharose 10%)	**2.08**	**0.923**
SA2 (filled with saccharose 20%)	**1.83**	**0.967**
SA4 (filled with saccharose 40%)	**1.68**	**0.941**
LS (filled with water, lower surface, 1/3)	**2.36**	**0.969**
LV (filled with water, lower volume, 0.5l)	**4.33**	**0.970**

The results show a sharp reduction in gas transmission through all the filled bottles, in comparison with the empty one; therefore the uselessness of the oxygen permeability determinations has been demonstrated for empty containers, which cannot be used in modelling the shelf-life of a liquid bottled, being these figures definitely far from the real oxygen accumulation rate. The permeability values are quite similar for all the filled bottles but those containing the sugar solutions (which have lower oxygen solubility) showed the lowest values. Varying the ratio between the permeable surface of the bottle and the volume of water contained, some interesting results were obtained: increasing this ratio (LV samples, i.e. half volume of water in the same bottle), the oxygen accumulation rate seems to accelerate, doubling the amount of oxygen dissolved in one day but, when a large proportion of the bottle walls was masked by a thick aluminium layer (i.e. decreasing the ratio with the same volume of liquid, LS), the oxygen transmission recorded was approximately the same as for the normal bottle filled with distilled water.

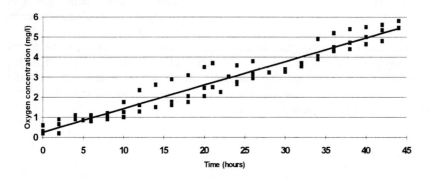

Figure 2 *Oxygen accumulation in a OPP bottle, filled with distilled water.*

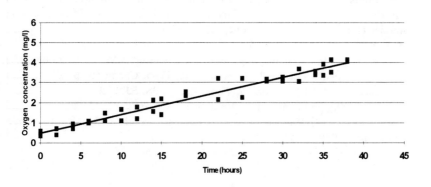

Figure 3 *Oxygen accumulation in a OPP bottle, filled with 40% saccharose*

These apparently conflicting results can be explained only considering the solubility and the diffusion rate of the gas in the medium as the prevalent aspects. The gas permeability constant is generally defined as the product of solubility and diffusion constants [7] but the determination of oxygen diffusion rate in foods, however, is more complex than oxygen solubility and data available is rather scarce [5,8]. An overall evaluation of our results leads to the emphasis of the role of gas solubility in the medium as the most important limiting factor for a forecasting of oxygen accumulation into a liquid packed in such a permeable bottle.

4 CONCLUSIONS

The results indicate that a medium barrier polymer such as polypropylene could find interesting applications in liquid packaging, provided that the oxygen tolerance of the medium is known and quite high, regardless of the high values of oxygen transmission rate, measured by conventional way.

The elimination of 3/4 of the permeable surface of the *OPP* container led to the same oxygen accumulation per volume, demonstrating the modest importance of the barrier properties of the container. The oxygen solubility in food, therefore, is the only factor that can be taken into account in selecting this type of bottle for the conditioning of liquid food or beverages with low sensitivity to oxygen. This fact also means that any size and shape of bottle can be designed and used provided that the medium can dissolve a high volume of oxygen without detrimental effects from the hygienic or sensorial point of view.

References

1. T.P. Labuza, "Kinetics of lipid oxidation in foods", T.E. Furia (Ed.), CRC Press, Boca Raton, USA,1971, p. 355.
2. E.N. Frankel, "Chemistry of autoxidation: mechanism, products, and flavour significance", in "Flavour chemistry of fats and oils", D.B. Mind and T.H. Smouse (Eds.), AOCS, New York, USA, 1975, p. 1.
3. M. Karel, *Food Technol.*, **9**, 50.
4. K. Eichner, "Food Packaging and Preservation", M. Mathlouti (Ed.), Elsevier Applied Science Pub. Barking, UK, 1986, Chapter 5, p. 67.
5. G.D. Sadler and P.E. Nelson, *J. Food Sci.*, 1988, **53**, 873.
6. P. Fava, L. Piergiovanni and G. Volonterio, *Vini d'Italia*, **33**, 27.
7. R.M. Barrer, "Diffusion in and through solids", University Press, Cambridge, 1959, p. 411.
8. C.S. Ho, L.K. Ju and C.T. Ho, *Biotechnol. Bioeng.*, **28**, 1086.

"Research supported by National Research Council of Italy, Special Project RAISA, Sub-project N. 4, Paper N. 1493."

Oxygen Permeability at High Temperatures and Relative Humidities

L. Axelson-Larsson

PACKFORSK, BOX 9, 16493 KISTA, SWEDEN

1 INTRODUCTION

All polymer materials are to some extent permeable to oxygen and the permeability increases with increasing temperature. As the temperature drops to the previous level, the permeability in many cases does the same. When packages are retorted, the material will be exposed to both high temperature and high relative humidity. This will strongly affect the barrier of some materials, e.g. ethylene vinyl alcohol (EVOH) and polyamide (PA), even after the temperature has dropped to the previous level, since the water will act as a plasticizer. The effect of the barrier is optimal immediately after sterilization, that is in the beginning of the storage and distribution. The permeability will vary with the relative humidity.

Generally, EVOH-materials are produced as multilayer materials consisting of at least five layers. The outer layers are generally PE, PP or PET, then there is an adhesive layer and in the middle the EVOH layer is placed. This structure will assure a very good oxygen- and water vapour barrier as long as the EVOH layer is dry. But when the package is retorted, the barrier of the outer layer against moisture will be lowered because of the high temperature used in the retort. This will give a higher moisture content in the EVOH layer and reduces its oxygen barrier. As the temperature drops after the retorting process, the barrier of the outer layer will go back to normal, which means that water will be trapped in the EVOH layer.

During storage the moisture content in the EVOH layer is decreasing. If, as in most cases, we have a high relative humidity in the package, the moisture content is at first the same or higher than the moisture content at equilibrium. So in this case moisture is transported from the EVOH layer and out through the outer layer. When we have reached the point, where the moisture content in the EVOH layer is the same or lower than the humidity in the environment, moisture will be transported into the EVOH layer from the inside of the package and then diffuse out through the outer layer.

2 THEORY

EVOH is a semicrystalline polymer and the permeation takes place in the amorphous regions. Permeability is an activated process and permeability as a function of temperature can be described by an Arrhenius equation:

$$P = P_0 \cdot e^{(-E/RT)} \tag{1}$$

where

P = Permeability
P_0 = Constant
E = Activation energy
R = Molar gas constant
T = Temperature

If the temperature is increased the permeability will increase, because the free volume in the material will increase. A common rule of thumb is that the permeability increases 30-50 % at a rise in temperature of 5°C.

Plots of lnP versus 1/T are linear, but the slope is changed, when the polymer reaches a transition temperature.

3 EXPERIMENT

The aim of the project was to measure the permeability before and after the retort process and to determine the permeability at retort conditions, see Table 1.

The studied materials were different combinations of PP/EVOH/PP and PP/PA/PP, see Table 2.

Table 1 *Retort Process at 121°C*

Heating up time	20 min
Retort time	20 min
Cooling and emptying	15 min

Table 2 *Materials Specifications*

Material	Composition	Thickness [m]	T_g after retort °C
A	PP/EVOH(32%)/PP	90/35/165	37
B	PP/EVOH(32%)/PP	90/35/505	37
C	PP/EVOH(32%)/PP	100/15/335	37
D	PP/EVOH(44%)/PP	234/29/242	38
E	PP/PA6/PP	131/66/134	36
F	PP/PA6/PP	179/85/202	

3.1 Permeability Measurements Made Before and After Retorting:

Permeability measurements were carried out in a Mocon Ox-tran at 23°C. The mean value of the relative humidity in the test cells were calculated to 60% RH [1].

3. 2 Determining the Permeability at Retort Conditions

Measurements of permeability were carried out at 45°C and 50°C. lnP = f(1/T) gives the oxygen transmission rate at other temperatures.

4 RESULTS

Figure 1 shows the measured permeability of the four PP/EVOH/PP materials after retorting. The barrier of the material D is thinner compared to A and B, but directly after retorting the barrier is better. One reason for this is the higher concentration of ethylene and another reason is that materials A and B have thinner outer layers which makes the moisture flow into the EVOH layer more easily during the retort process. We have to wait 70 to 80 hours before materials A and B catch up. However, the thinner outer layer also results in materials A and B "drying out" faster. The permeability after three days is between 50 and 100 times higher than before retorting.

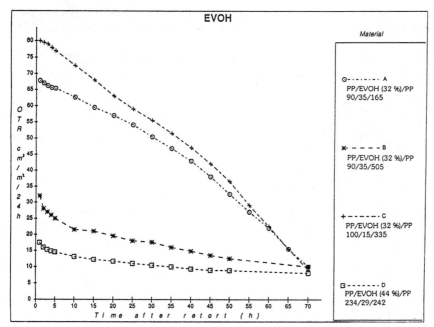

Figure 1 *Permeability after retorting EVOH material*

Figure 2 shows the permeability measurements for the PP/PA6/PP materials after retorting. In this case the permeability 3 days after retorting is about 2-4 times higher than before retorting

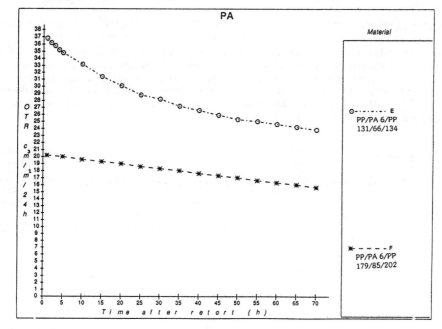

Figure 2 *Permeability of PA material*

The results from the permeability measurements at two temperatures above T_g are shown in Figure 3. This Figure also shows the extrapolation to the retorting temperature 121°C.

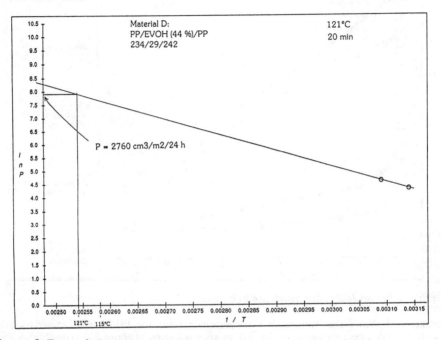

Figure 3 *Extrapolation to retort temperature*

5 CONCLUSION

The permeability of moisture sensitive materials after retorting is much higher than before retorting a long time after the processing. The extrapolation method is an easy way to determine the permeability at retort conditions, compared to mathematical models. The method has to be validated by measuring the permeability at more than two temperatures above T_g. It should also be done using equipment with more accurate temperature regulations than in the Mocon Ox-Tran [1].

References

1.   L. Axelson-Larsson, *Packaging Technology and Science*, 1992, **15**, 297-306.

Interaction Phenomena and Barrier Properties of Plastic Packaging Material in Relation to Foodstuff

B. Lindberg

THE DANISH PACKAGING AND TRANSPORTATION RESEARCH INSTITUTE,
DANISH TECHNOLOGICAL INSTITUTE, GREGERSENSVEJ, PO BOX 141, DK-2630
TAASTRUP, DENMARK

1 INTRODUCTION

When talking about interaction between plastic packaging material and foodstuff (incl. drinks) the migration phenomenon is a main topic. For the packaging expert/scientist migration is the unwanted transportation of mobile molecules from the plastic material out to the foodstuff. An important task is to minimize this migration due to health aspects. Thus many countries and now also EU (European Union) have regulations about limited (maximum) migration values for plastic packaging material intended to come into contact with foodstuff. We here talk about plastics categorized as "food quality".

In the developed countries there have for a long time existed laws and regulations about migration such as BGA in Germany, Warenwet in Holland, FDA in U.S.A and now also a long list of EU-directives[1]. The regulations deals both with total migration and specific migration (for a specific chemical component). Concerning total migration, the maximum accepted migration today in the EU-directive (90/128) is set to 10 mg/dm^2. Thus the demand on the packaging material is that the amount of components coming out from the plastic may not exceed the specified migration values.

Another type of interaction is what packaging experts call barrier effect. In this case the effect normally is to hinder small molecules from penetrating and passing through the plastic packaging material. Concerning protection of foodstuff and obtaining long "shelf life" good barrier effect, especially in relation to oxygen, is important. More principally, barrier effect is also wanted in relation to microbiological attack and radiation (e.g. protection against deteriorating sun-light).

The described interaction phenomena are presented in Figure 1

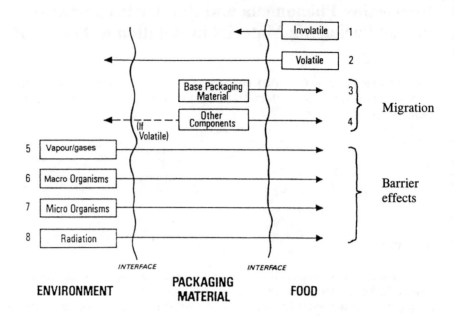

Figure 1 *Components for interactions with: environment/packaging material/food*

2 COMPOSITION OF PLASTICS AND MIGRATION

When studying interaction-phenomena, such as migration and barrier effects, it is of interest to look a little closer at the composition and the material science of plastic. Technical plastic material can be considered to consist of plain polymer plus additives;

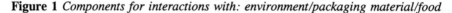

Plastic = Polymer(s) + Additives

Some packaging plastics are used with almost no extra additives, while other types can have rather high amounts of additives.

Already in the production of the raw material (the polymer), called polymerisation, certain process additives must be used like catalysts, initiators, retarders, emulsifiers, stabilizers etc. These additives will normally be encapsulated in the polymer as impurities.

The producer of packaging materials and products (film, trays, bottles etc.), also called the packaging converter, need to add further specific components (additives) to the "pure" polymer to obtain certain properties. These properties can be related to production conditions or to give the ready packaging certain character e.g. colour (Figure 2).

Polymerisation process (pure polymer)	-	catalysts
	-	initiators
	-	retarders
	-	emulsifiers
After polymerisation/ before processing (technical plastic)	-	antioxidants
	-	stabilizers (heat)
	-	plasticizers
	-	colour pigments
Processing/converting (packaging plastic)	-	slip agent
	-	antistatic agent
	-	anti blocking
	-	UV-absorbers
	-	pigments/fillers
	-	anti-dew
	-	foaming agents
	-	antioxidants
	-	biocides

Figure 2 *Examples of additives used for production of plastic packagings.*

The different types of additives commonly used in production of packaging plastic are shortly reviewed below[2].

Catalysts and initiators are used to start the polymerisation reaction between monomer molecules to build up macro molecules (polymers).

Antioxidants protect the polymer from heat degradation e.g. under thermoforming of the material.

UV-stabilizers increase the resistance against UV-light degradation (e.g. returnable plastic bottles).

Plasticizers reduce stiffness and make the material very elastic and give film good "cling" properties (self-adhesion).

Anti-blocking prevent the plastic from blocking together (e.g. used for production of plastic bags).

Antistatic agents are used to prevent building up of static electricity in the material, which may cause unpleasant "touch" and attraction of dust to the surface.

Anti-condense prevent formation of condense droplets (dew) on transparent film and packaging windows.

Fillers are used to reduce material cost, obtain more bulk and can also increase barrier properties.

Pigments and colorants give the material the wanted colour and give UV and visible light protection.

Foaming agents are chemical components used to produce expanded plastic (e.g. EPS for fish and meat).

Based on the above mentioned components used in plastic material we can formulate a "master" formula for a packaging plastic containing some typical additives (Figure 3).

Polymer	Conc. % ~ 98
Antioxidants (phenols, phosphites, thioester)	0.02-1
UV-stabilizers (benzotriazols, benzophenones)	0,2-0,5
Softening agents, plasticizers (Sn-stearic amide, phthalates, adipates)	< 20 >
Antistatics (fatty acid ester with hydroxy or amine groups)	< 1
Slip agents (erucamide)	0,02-0,1
Neutralizer (Ca-stearate)	0,01-0,2
Antiblock agents (silica)	0,05-2
Crystallizing agents (Na-benzoate)	< 0,1
Monomers (VCM, styrene, acrylonitrile)	very low
By-products from processing (aldehydes, ketones, acids)	?
Solvents from laminates and printing	?

Figure 3 *Possible components in plastic material (for packaging usage)*

A lot of different types of plastic are used for production of packaging products. The dominating group is the thermoplastic polymers and the most important are given below:

Polyolefines:	Polyethylene	LDPE
		LLDPE
		MDPE
		ionomers
		copolymers (EVAc, EBA)
	Polypropylene	PP
Vinyls.	Polystyrene	PS
	Polyvinylacetate	PAc
	Polyvinylchloride	PVC
	Polyvinylidene chloride	PVDC
	Polyacryl nitrile	PAN
Esters:	Polyethylene terephthalate	PET
	Polycarbonate	PC
Amides:	Polyamides (Nylon)	PA6, PA66, PA11

As mentioned before different chemicals are used to initiate or start the chemical reactions to produce the polymers. Perhaps not all monomer react and form polymers.

In addition rest monomers exist in the polymer, especially concerning PAN and PVC, where there are very restricted regulations about presence of rest monomers. The EU-directive 90/128 says: SML acrylonitrile = 0,020 mg/kg (detection limit) and for vinyl chloride
SML = 0,01 mg/kg. Here SML = Specific Migration Limit for the specific monomer.

3 PLASTIC BARRIER PROPERTIES

Barrier properties of the packaging have influence on the shelf life of food products and on food quality[3]. Knowing the barrier data for oxygen transmission (permeability) through the packaging, the shelf life for food products with known levels of maximum oxygen up-take may be theoretically calculated. Different food products have different levels of maximum oxygen gain or up-take in relation to shelf life. Some values have been presented by Salame[4] giving maximum oxygen gains of various foodstuffs and beverages before spoilage in one year storage at 23°C. Some examples are given below:

1.	Retorted food (meat, fish)	1 - 5 ppm
2.	Beer and wine	1 - 5 ppm
3.	Fruits (in can)	5 - 15 ppm
4.	Fruit juices	10 - 40 ppm
5.	Oils and fats	50 - 200 ppm

Based on this knowledge the level of barrier effect for a specific packaging situation can be calculated.

Today there are advanced instruments for quick and easy measuring of oxygen and water permeability even for plastic packaging with very low permeability (down to 0,1 cm^3/m^2 x 24 h x atm). These instruments are based on registration of test gas molecules by use of e.g. accurate and sensitive detectors (for oxygen colormetric analysis and for water IR-analysis). Standardized methods exist for measuring permeability such as ASTM D-3985 for oxygen and ASTM F-372 for water vapour.

Traditionally high barrier effect for plastic packaging has been achieved by using PVDC, EVOH or aluminium as barriers in thin layer in combination with other base polymer (such as PE, PP, PET). Due to very high prices and also due to technical aspects PVDC and EVOH are not used as the only polymer in packagings. PVDC has been criticized for causing environment problems (produces hydrochloric acid and can also be a source for formation of dioxin during incineration) and EVOH have some disadvantages as the good oxygen barrier effect is weakened under influence of high humidity. Concerning aluminium also negative environmental figures have been put forward and another disadvantages is that the product can not be seen in an Al-packaging (Al-foil or metallized Al).

In the work of developing new types of "High Tech" barrier films have been developed and newly introduced on the market[5]. One interesting type is plastic coated with a barrier film of an inorganic metal oxide like silicon oxide.

This type of coated plastic is often referred to as "flexible glass". Silicon oxides used for barrier application are referred to as SiO_x. This is because the high vacuum evaporation process yields a mix of silicon mono and dioxide.

The X-values range between 1.5 - 1.8. The starting material for the evaporation process is usually silicon monoxide, SiO^6. Low X-values give the best oxygen barrier effect, but also the highest tendency for yellowing. A good compromise with respect to these two contradictional properties is to choose a X-value about 1.7[7]. The coating technique using high vacuum deposition of the inorganic material may be used on flexible plastic films or on plastic bottles. The inorganic coating layer have good barrier effect against diffusion of gases (oxygen), water, aroma etc. Also other inorganic material such as Al_2O_3 has been suggested and studied as barrier coating[5].

Further more SiO_X is very inert and would not cause problems with respect to the EU Directive on migration (90/128 EEC). Oxygen permeability data for barrier laminate film based on SiO_X as barrier layer compared to more traditional barrier laminates are presented in (Figure 4).

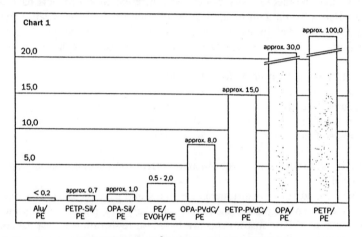

Figure 4 *Oxygen permeability (ccm/m^2 x 24 h x bar) of different material combinations.*
NB! "Sil" stands for silicon oxide coating.
Technical Information "4P Verpackung", Germany.

Today silicon oxide barrier is normally offered as a laminate with the very thin SiO_X-layer protected between plastic layers. Relative complex laminate systems are formed as also printing and glueing processes are involved (Figure 5).

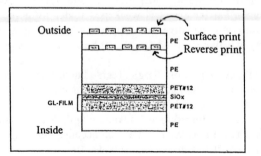

Figure 5 *Principle for built-up of SiO_X-barrier laminate.*

4 ENVIRONMENTAL ASPECTS

The new EU draft on packaging waste[8] is putting pressure on traditional laminates build-up by different materials, as such products may be difficult to recycle. The SiO_x-layer on this new type of barrier packaging is extremely thin (around 0.1 micron) and can therefore be considered to be an integrated part of the coated plastic material itself and be recycled as a homogenous material.

Ecobalances, or life-cycle analyses (LCA) are becoming increasingly important as a mean of analysing the impact of manufactured goods on the environment. LCA should provide a quantitative measure of the natural resources consumed throughout the life of a product, from extraction of the raw materials from which it is made, through its useful life, to its destruction. New packagings cannot be developed without paying proper attention to the environmental aspects. Many countries have introduced taxes or fees on packagings in relation to their impact on the environment: The Green Dot (Der Grüne Punkt) in Germany and Austria, and The Eco-Emballages in France etc.

5 CONCLUSIONS

The classical interaction phenomena concerning plastic packagings like migration and permeability (barrier effect) are still of scientific and technical importance, especially in relation to the development and introduction of new packaging materials and products due to the "green" tendency in the society.

References

1. F. Lox. Packaging and Ecology, Pira International, Leatherhead, 1992, p. 263.
2. Modern Plastics Encyclopedia, Mc Graw-Hill Inc. 1989, p. 143.
3. B. Lindberg, Barriere egenskaber og plastemballager til levnedsmidler. *Plus Process* 1992 no. 7/8, p. 12 and no. 9, p. 20.
4. M. Salame. The use of low permeation thermoplastics in food and beverage packaging. *ACS, Div. of Organic Coatings and Plastics Chem.* 1974, 24 (1), p. 516.
5. Shives. Developments in Barrier Technology. A literature review. Pira International, Leatherhead, 1992.
6. P. Maplestone. Barrier film coatings add performance options. *Modern Plastics Int.* 1992, 22, 8, p. 22.
7. J. Neutwig. SiO_x-Barriere-Folien. *Neue Verpackung*, 1993, 5, p. 45.
8. Proposal for a council directive on Packaging and Packaging Waste; Com 1993, 416 final - SYN 436, Brussels.

A NIR Reflectance Method to Measure Water Absorption in Paper

C. Bizet, S. Desobry, and J. Hardy

LABORATOIRE DE PHYSICOCHIMIE ET GÉNIE ALIMENTAIRES, ECOLE NATIONALE SUPÉRIEURE D'AGRONOMIE ET DES INDUSTRIES ALIMENTAIRES (ENSAIA) INSTITUT NATIONAL POLYTECHNIQUE DE LORRAINE (INPL), 2, AVENUE DE LA FORÊT DE HAYE - BP 172, 54505 VANDOEUVRE-LÈS-NANCY CEDEX, FRANCE

1 INTRODUCTION

Highly calendered papers are often used for the packaging of soft cheeses. They cannot be used alone because of a too high water vapor permeability but are often one-side coated or combined with other packaging materials like oriented polypropylene (OPP), polyamide (PA), polyester (PET), polyethylene (PE),...[1,2]. Packaging materials meet requirements as far as safety, mechanical and transfer properties are concerned. Because legislation is very strict with packaging materials in contact with food products, producers need to have reliable control methods[3]. Several parameters are controlled during the paper production process and on the final product like optical parameters -colour, haze, glaze, light transmission...-, mechanical parameters, physical parameters -porosity, rugosity, weight, thickness,...- or water behaviour [4]. As far as water is concerned several tests exist like the measure of maximum quantity of water absorbed -COBB test-, water vapor permeability or total water desorption rate[5]. These measures are static ones and they give no information about kinetic profiles of water absorption by paper when it is laid on food products. Desobry[6] showed that it could be interesting to know dynamic parameters of water absorption as it influences cheese storage quality. Water content can be measured by several techniques : direct weighing, dielectric properties or near infrared (NIR) reflectance properties[4].

NIR reflectance method has already been used in many applications and a very complete work has recently been published on this subject[7]. Several components can be analysed like water, proteins, lipids, carbohydrates,...In the packaging area it is already used for the analysis of packaging laminates[8] or cellulose derivatives[9].

The aim of this study was to measure water absorption of paper -water content versus time- in contact with a simulated high moisture food made of agar gel and therefore we investigated a NIR reflectance method.

2 MATERIALS AND METHODS

2.1 Simulated Food and Paper Samples

A reproducible high moisture food similar in shape and size -110 mm diameter x 14 mm thick- to Camembert cheese was simulated by using a 2% w/w agar gel. A simple kraft paper containing 41 g.m^{-2} of paper fibers was used. Samples were placed in dry regulated hygrometry boxes during several days. Hygrometry was regulated thanks to a supersaturated LiCl solution.

2.2 NIR Reflectance Data

When a material receives radiation, a part of it is transmitted. However when the material is thin enough, which was not our case, a part is absorbed if the frequency is the same as the one of the molecule vibration and the last part is reflected. It has been proved[10] that log(R'/R) is directly related to the concentration of water, R' being the reflectance at a reference wavelength and R the reflectance at a measure wavelength. In our case R' was a combination of 2, R'_1 and R'_2 that are supposed to limit the influence of some parameters such as colour, granulometry, other components or surface state. R' remains constant with water variations so log(1/R) carries all informations.

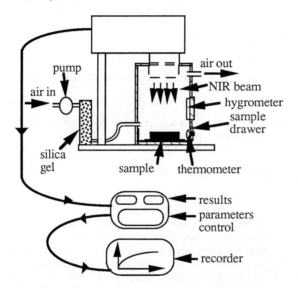

Figure 1 *Arrangement for the measure of surface cheese water content by a NIR reflectance method*

A Quadra Beam 6000 (Moisture System Inc., Hopkinton, U.S.A.) was used in the NIR region of the spectra. It was equipped with a 3-filters wheel. Two of them carried reference wavelengths -1800 nm and 2120 nm- where water does not absorb the radiation. The last one carried the measure wavelength -1940 nm- where water absorbs the radiation.

Measurements were made at 11 °C ±2°C and 11% relative humidity. Both were regulated and controlled as described in Figure 1.

2.3 Experiments

Fifty samples were placed on agar gel under the NIR beam and left during different times. Log(R'/R) was directly recorded and we measured water content of paper by removing it from agar gel and weighing. Data couples [water content by weighing ; log(R'/R)] were introduced in the Moisture Meter and calibration parameters were calculated. The calibration was tested with 25 more samples by comparing water content measured by weighing versus water content by direct NIR reflectance reading. Then a sample was left on agar gel and water content was recorded continuously in order to follow water absorption through paper.

2.4 Finite Element Method

To modelise experimental kinetic profile, we used the finite element method. The principle was that we divided the solid -paper sample- in 'n' slices, each Δx thick. The hypothesis that the transfer was in one direction -from agar gel through paper to ambient air- was made and we wrote the transfer equation in the nth slice using Fick's law :

$$C_n^{t+\Delta t} = \frac{D\Delta t}{(\Delta x)^2} \cdot (C_{n+1}^t + [\frac{(\Delta x)^2}{D\Delta t} - 2] \cdot C_n^t + C_{n-1}^t) \qquad (1)$$

where D : diffusion coefficient
Δt : period of time
Δx : thickness of one slice

C_n^t : water concentration in the nth slice at the time t
In order not to be incompatible with the second law of thermodynamics, we must have $\frac{(\Delta x)^2}{D(\Delta x)} \geq 2$. To computerize equation (1) we took $\frac{(\Delta x)^2}{D(\Delta x)} = 2$. We considered that the first slice absorbed immediately the maximum water quantity. By computer treatments we could find a curve that fitted well with experimental points.

3 RESULTS AND DISCUSSION

3.1 Calibration

In the Figure 2, each point was an average of 5 paper samples with CV<5%. It is possible to find a good correlation between NIR reflectance data and water content obtained by weighing. Several authors[7] have pointed out that one of the main problems of such a method is to find the reference method. It does not matter how difficult, expensive or long this one is as long as it is reliable and precise. In our case, a simple measure of different weights -before and after water absorption- could be considered as a reference method.

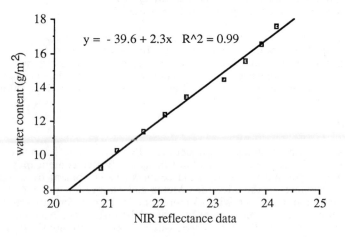

Figure 2 *Calibration : water content (g/m^2) absorbed by paper measured by weighing versus NIR reflectance of paper surface*

Concerning the depth of penetration of NIR radiations, it varies from some microns to millimeters according to the material and the number of molecules met[11,12]. In our case, it is possible that the radiations penetrated more than the paper into the agar gel, but we considered that it was approximately the same for each [paper+agar] sample. A mistake was made but it was cancelled by the calibration.

3.2 Validation

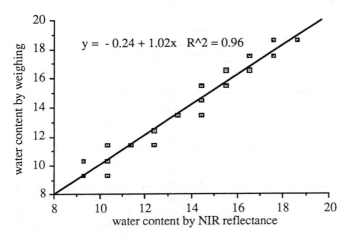

Figure 3 *Validation : water content of paper measured by weighing versus water content read by NIR reflectance method*

Figure 3 shows that there was a good prediction of water content absorbed by 25 new paper samples. A perfect calibration would have given a straight line with an equation 'y=x'. In our case the equation is a 'y=ax+b' one, but we can notice that a is close to 1 and b close to 0 when compared to the whole scale.

3.3 Kinetic Profiles

A complete kinetic profile of water absorption by paper was recorded (Figure 4) and we notice that the curve presents two types. First a very fast increase of water content then a decrease of water absorption speed. The curve tends to be asymptotic and would reach the maximum water quantity that can be absorbed by paper for an infinite time. It seems difficult to make a strict exploitation of this curve and that is the reason why we tried to find a model that would fit to experimental points. We investigated the finite element method as described in 2.5 and we found out that the behaviour of the slice n°2 represented the reality when parameters were chosen as follows :

$$\Delta t = 11.7 \text{ s and } (\Delta x)^2/D = 23.4 \text{ s}^{-1}$$

We must be aware that what is measured by NIR reflectance is not only what happens in slice n°2 but that the behaviour of water in slice n°2 is a good approach of the reality measured by NIR reflectance.

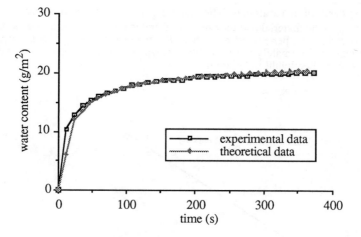

Figure 4 *Kinetic profile : water content of paper measured by NIR reflectance versus time*

By this method we were able to measure and calculate several parameters characteristic of this kraft paper like the maximum water quantity absorbed which was 23.9 $g.m^{-2}$ for kraft paper, the time to absorb for example 80% of this quantity which was 193 s or the mean speed of water absorption for 80% of maximum water absorbed which was 0.1 $g.m^{-2}.s^{-1}$. These numbers would certainly be very different for any other kind of paper. Moreover these parameters probably depend on the technological parameters of paper production as well as structural parameters like paper porosity, fibers entanglement,...

4 CONCLUSION

A new near infrared reflectance method for the measure of water absorption by paper was investigated and validated in the case of a simple kraft paper laid on a high moisture simulated food. It was possible to relate water content of paper sample with NIR reflectance data and we proved the reliability of the calibration. We perfected the exploitation of kinetic profile of water absorption by the finite element method. We found out that it was possible to get several parameters from the curve that would give informations on paper behaviour. We possess now a good tool to know water absorption of paper and we should be able to compare different kinds of paper as well as coated papers and papers combined with other materials. We could also relate some of the parameters we measured with other properties of paper like porosity, COBB test,...

Acknowlegements

We would like to thank VENTHENAT S.A., part of CMB Packaging, for their financial and technical support.

References

1.   G. Stehle, 'Le fromage', A. ECK, Lavoisier, Paris, 1984, p. 367.
2.   FIL-IDF, Packaging of butter, soft cheese and fresh cheese, 1987, **214**, p.12.
3.   R. Souverain, 'Laits et produits laitiers', Tec et Doc, Lavoisier, Paris, 1986, p.415.

4.   P. Vallette and C. de Choudens, 'Le bois, la pâte, la papier', Centre Technique de l'Industrie des Papiers, Cartons et Celluloses, Grenoble, 1989.
5.   S. Desobry, Thèse de Doctorat, Institut National Polytechnique de Lorraine, 1991.
6.   S. Desobry and J. Hardy, *Int. J. Food Sci. Technol.*, 1993, **28**, 347.
7.   B.G. Osborne, T. Fearn and P.H. Hind, 'Practical NIR spectroscopy with applications in food and beverage analysis', Longman Scientific and Technical, Wiley & Sons, New-York, 2nd ed., 1993.
8.   A.M.C. Davies, A. Grant, G.M. Gavrel and R.V. Steeper, *Analyst*, 1985, **110**, 1034.
9.   W.I. Kaye, 'The Encyclopedia of Spectroscopy', Clark G.L., Reinhold, New-York, 1960, p.409.
10. P. Kubelka and F. Munk, *Zeitschrift für technische Physik*, 1931, **12**, 593.
11. B.G. Osborne, *J. Fd. Technol.*, 1981, **16**, 13.
12. D. Bertrand, P.R. Robert, F. Evin, F. Le Corre and M. Pinel, Les Cahiers de l'ENSBANA, Tec et Doc, Lavoisier, Paris, 1990, p.95.

A Static Head Space Gas Chromatography Method to Measure Plastics Permeability to Ink Solvents

S. Desobry, C. Bizet, M.N. Maucourt, G. Novak, and J. Hardy

LABORATOIRE DE PHYSICOCHIMIE ET GÉNIE ALIMENTAIRES, ECOLE NATIONALE SUPÉRIEURE D'AGRONOMIE ET D'INDUSTRIES ALIMENTAIRES (ENSAIA), 2, AVENUE DE LA FORÊT DE HAYE, BP 172, 54505 VANDOEUVRE-LÈS-NANCY CEDEX, FRANCE

1 INTRODUCTION

Residual solvents in wrapping materials cause toxicological problems. European standards impose a limited amount of residual solvents in packaging materials and in foods. To reduce solvent migration into packaging materials during printing and into food during storage, the permeability, P, and apparent solvent diffusivity, D, in the packaging material must be known. With low D values, solvents migrate slowly in the packaging material and the quantity of residual solvents in packaging material is limited by drying. Great D values lead to easy migration and expensive drying.

Knowledge of permeability and diffusivity could help industries to choose the solvent well adapted to the packaging material. Some authors[1,2] presented interesting methods to calculate organic vapor transmission rate through film, using the Gilbert-Pegaz cell. However, it was recently observed that, in dynamic methods, the nitrogen circulation in the cell accelerated the vapor transfer through film. Permeation was increased.

Recently, Thalmann[3] studied the sorption, desorption and retention of solvents in plastics, but gave only few diffusivity values.

The aim of this paper was to develop and validate a static chromatographic method to determine permeability and diffusivity of solvents through flexible packaging materials. To test the method, a polyethylene film and two solvents - 1-propanol and 2-propanol - were chosen.

2 MATERIALS AND METHODS

2.1 Solvents

We chose two organic solvents used in packaging industry in ink and varnish, i.e. 1-propanol and 2 propanol. They are chemically similar but their molecular volumes are different and modified diffusional properties. Density of propanol is 0.8.

2.2 Packaging Material

A 50 μm thick Polyethylene film was tested. Polyethylene (PE) represents 50% of the plastic used in food packaging industry.

2.3 Static Head Space Gas Chromatography (SHSGC)

A Perkin-Elmer SHSGC (PE 8500 - HS 6) was used with a aluminium cell developed in our departement (Figure 1). The SHSGC parameters are given in Table 1.

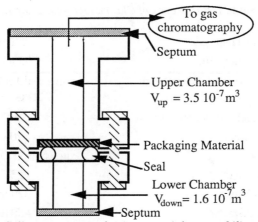

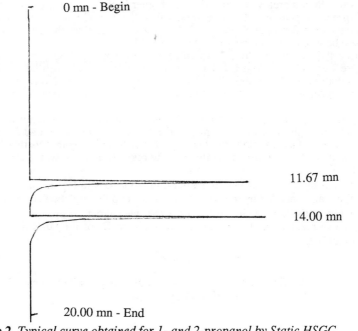

Figure 1 *Cell to measure packaging material permeability to solvents*

Calibration was accomplished using an unpermeable 20 µm thick aluminium sheet as packaging material. Zero to 0.6 µl of solvent were deposited in the upper chamber. After 10 minutes at 70°C, the air in the upper chamber was agitated during 15 seconds and one sampling was automatically performed and injected in the column. Lowest temperature was chosen to ensure good vaporization of propanol. A typical chromatographic profile is presented in Figure 2.

To determine PE film permeability, 15 µl of the solvent was deposited in the lower chamber to obtain saturated air. At 70°C, solvent migrated through the film. After a time t, the upper air composition was analysed as previously.

Figure 2 *Typical curve obtained for 1- and 2-propanol by Static HSGC*

Table 1 *Parameters of Chromatographic Analysis*

Initial temperature	140 °C
Iso-temperature time	2 min
Rate of Increasing Temperature	3°C/min
Final temperature	180 °C
Injector and detector temperature	230°C
Cell temperature	70°C
Column type	Porous Layer Open Tubular
Column length	25 m
Internal diameter	320 μm
External diameter	450 μm
Film thickness	10 μm

3 RESULTS AND DISCUSSION

To calculate the diffusional parameters of propanol through PE film, we used the calibration curves presented in Figure 3. Peak height (Figure 3-a) and area (Figure 3-b) were analysed.

First, it appeared that correlation coefficients between peaks height and solvent quantities deposited in upper chamber were the best.

Secondly, for the same quantity of 1-propanol and 2-propanol in the upper chamber, peaks height and area were considerably higher in the case of 2-propanol. The height of the 2-propanol peak was 1.7 time greater than the one of 1-propanol. On the other hand, the area of the 2-propanol peak was 2.0 times greater than the one of 1-propanol. The relation between area and height of peaks is not linear.

These two points could be explained by the peaks tailing, observed on the graph (Figure 2), resulting from diffusion of the solvent in the capillary column. As area determination is difficult because of peak tailing and as correlation coefficient is the best one, the permeability and diffusivity of the solvents were calculated from the height of the peaks.

A linear relationship was obtained between time, t, and propanol amount that have migrated through PE film. In Figure 4, the curves of 1-propanol and 2-propanol seemed to be equal. In fact, by plotting 'quantity of transferred propanol vs time' (Figure 5), the permeability, P, calculated from slope of the regression draw, was greater for 1-propanol (Table 2).

Diffusion time lag, τ, is the intercept on the time axis in Figures 4 and 5. This time lag is needed by the solvent to cross the polyethylene film. Some authors have also observed the time lag in the case of barrier packaging[1].

From these results, we determined the diffusivity, D, of propanol through PE film, by using Fick's law[4].

$$D = \frac{e^2}{6\,\tau} \qquad (1)$$

where : τ = diffusion time lag of propanol through polyethylene (s)
             e = thickness of PE film (e = 50 μm)

3a -

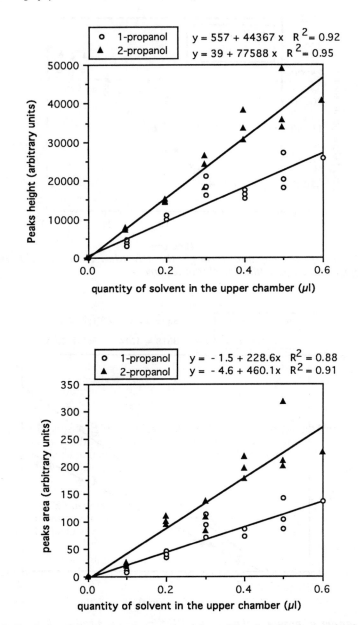

3b -

Figure 3: *Static head space gas chromatography calibration. Peaks height (3a) and area (3b) vs quantity of propanol deposited in the upper chamber of the cell.*

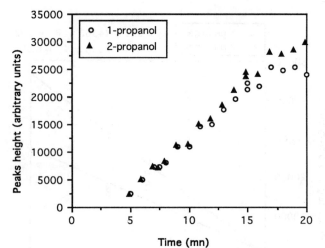

Figure 4 *Transfer of propanol through the PE film. Peaks height given by chromatographic analysis as a function of transfer time.*

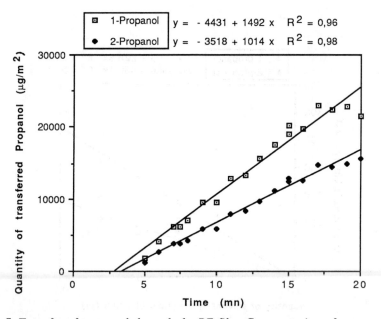

Figure 5 *Transfer of propanol through the PE film. Concentration of propanol in the upper chamber of the cell as a function of transfer time.*

τ and D are given in Table 2. Diffusivity of 1-propanol was greater than diffusivity for 2-propanol. This certainly resulted from the chemical structure of both molecules. 1-propanol has a quasi-linear structure and migrated easily through PE. 2-propanol has a quasi-spherical structure and its molecular volume could limit permeability through the film.

Table 2 *Permeability of PE and Diffusivities of 1- and 2-Propanol through PE*

Solvent	Permeability $(g.m^{-1}.s^{-1}.atm^{-1})$	Diffusion time lag (s)	Diffusivity $(m^2.s^{-1})$
1-propanol	$12.4 \ 10^{-10}$	177	$2.35 \ 10^{-12}$
2-propanol	$8.4 \ 10^{-10}$	207	$2.01 \ 10^{-12}$

It seemed that industrials would use branched alcohols to limit diffusion in polymer and then to limit residual solvent content.

4 CONCLUSION

In this paper, we demonstrated the validity of the Static Head Space Gas Chromatography to determine the diffusivity of solvents and package permeability. This method is rapid and easy to use. It is now necessary to develop this method to other solvents and packaging materials used in food industry.

References

1.   M.G. Kontominas, *Sci. Aliment*, 1985, **5**, 321.
2.   E. Hatzidimitriu, S.G. Gilbert and G. Loukakis, *J. Food Sci.*, 1987, **52**, 472.
3.   W.R. Thalmann, *Packaging Technol. Sci.*, 1990, **3**, 67.
4.   S. Desobry and J. Hardy, *Int. J. Food Sci. Technol.*, 1993, **28**, 347.

Determination of Microholes in Food Packages

L. Axelson-Larsson[1] and E. Hurme[2]

[1]PACKFORSK, BOX 9, 16493 KISTA, SWEDEN
[2]VTT BIOTECHNOLOGY AND FOOD, RESEARCH LABORATORY, PO BOX 1500, FIN-02044 VTT, FINLAND

1 THE PRINCIPLE OF THE ELECTROLYTIC CONDUCTANCE METHOD

When a voltage is applied over two electrodes in an electrolytic cell, a current will run through the circuit and Ohm's law is valid. In accordance with Ohm's law, the current is proportional to the applied voltage according to:

$$I = G * U \tag{1}$$

where
I = Current (A)
G = Conductance (S)
U = Voltage (V)

In this case a package filled with electrolyte is placed in an electrolytic bath. If one electrode is positioned into the package and the other electrode is positioned into the electrolytic bath, the ion transport will be stopped provided that there are no holes through which the ions can pass.

The measured current and the applied voltage gives the total conductance of the cell. There are many factors more or less contributing to the conductance, for example conductance in the electrodes, polarization phenomena at the surface of the electrodes and conductance in the solution and in the microhole. In this case the conductance in the hole is demanded. Therefore calibration of the method is necessary. When the method is calibrated, the diameter of the hole can be calculated, by using the formula for electric conductivity .

$$d^2 / 1 = G * * 4 / \pi \tag{2}$$

where
d = Diameter [m]
1 – Length [m]
G = Conductance [S]
= Resistivity [$S^{-1} m^{-1}$]

According to this law the ratio of the square diameter and the length is proportional to the measured conductivity and the resistivity of the solution. The equipment Microhole Tester is based on this method and the development is reported by Axelson, Cavlin and Nordström in ref [1].

2 THE INSTRUMENT MICROHOLE TESTER

The instrument consists of an electrolytic bath, with stirring, temperature regulation and a portable power source connected to a programmed calculator. The instrument has been calibrated by testing several cups with well defined microholes. The smallest microholes made were 5μm in diameter. The programme calculates the diameter of the hole and calibration is carried out automatically.

2.1 Test Procedures

Before the actual testing starts the following procedure makes sure that the electrolyte is correctly mixed. The electrodes are placed in fixed positions in the electrolytic bath and the conductivity of the salt solution (1 % NaCl) is measured with AC at 0.5 V. The programme tells step by step how to do the testing and tells if the resistivity of the salt solution lies within a tolerable range.

1. The packages that are to be tested are filled with 1 % NaCl and positioned into the electrolytic bath. One electrode is positioned in the bath and the other electrode is positioned in the package.

2. The measuring starts by pressing the voltage button and DC at 10 V is running through the circuit. If there is no hole in the package, the current will be zero. However, if there is a hole the value of the current will be related to the hole size.

3. The read off value of the current is inserted into the programmed calculator and then the programme asks for an estimated hole length.

4. The diameter of the hole is given on the display.

3 FIELD OF APPLICATION

The method is suitable for testing any kind of package, like for example trays, cups, bottles and bags. However, a problem may arise while testing packages that contain an aluminium layer. In the case of plastic and aluminium laminated paper packages the current will be zero if there are no pinholes, holes in the seals or cracks in the plastic layer. If there is a hole in the seal and the plastic layer covers the aluminium layer, there is no problem in measuring the hole size. However, if the electrolyte gets in contact with the aluminium foil, for example through a microcrack in the inner plastic layer, the current signal will fluctuate very much. This is because the electrolyte in the package gets in direct contact with the aluminium layer which acts as an electric bridge between the two electrodes applied. The electrolyte will react chemically with the aluminium layer and this
results in gas formation. In this case the conductance will decrease because gas bubbles will hinder the ion transport. It is difficult to estimate the hole size in this case, due to the fact that the current fluctuates and does not go only through the electrolyte but also through the aluminium layer.

Surface agents might be needed for making the material wet by the electrolyte and making it possible for the electrolyte to penetrate very small holes. This must be tried out from case to case.

The Microhole Tester is an instrument for laboratory control of packages. It is a very suitable instrument for packaging design and has been used for many different kind of packages, not only sealed packages but also packages with screw caps. The tightness and integrity using different sealing techniques has been evaluated as a part of the development of a new package.

The Microhole Tester is a laboratory control method suitable for random sampling of packages. For a packaging line with high production speed and a hard quality limit it is not possible to use the equipment for statistical control, because of the testing time. Statistical control by leakage detection sytems are discussed by Axelson *et al* [2].

The Microhole Tester has been used in research programs for evaluation of leakage control methods. In the evelution study the Microhole Tester was used as a control instrument and dye-test was used for locating the microholes.

4 EXPERIMENTAL EVALUATION OF THE RELIABILITY

4.1 Influence of Testing Time

When the electrode is positioned in the package the current will generally almost immediately stabilize at a level proportional to the hole size. However, if the holes in the package are very small (less then about 20μm in diameter), salt crystals tend to precipitate and block the hole. Consequently the current drops. This will occur after a testing time of approximately 5 - 10 minutes. Thereforethe values should be read off after approximately 2 minutes. In the case of plastic laminated paper packages there might not be a hole through the package, but a microcrack in the inner plastic layer. In this case the current will be very small (10 -100 times lower than values obtained from measuring the current through capillaries with a diameter of 25 - 100 m and a length of 0.5 - 2 cm), but it will increase as the measuring continues due to the fact that paper will absorb water. However, still there is no problem if the values are read off in approximately 2 minutes.

4.2 Influence of the Content

The content may affect the testing time for example if the food is blocking the hole. The longer the package is placed in an electrolytic bath before testing, the shorter the actual testing time will take to reach a stable value. This must be tried out from case to case in order to get a reliable and stable value at shortest possible testing time.

Test 1

In tray packages made of HDPE and with a lid material of PE/PETP, microholes were made by pulling a tungsten thread through the seal*. 102 packages were empty and 40 packages were filled with steak. The packages were emptied and filled with electrolyte two days before testing.

Test 2

208 tray packages made of PP/EVOH/PP and with a lid material of PP/EVOH/PET were filled with mashed potatoes. Microholes were made in 104 of the packages by placing a tungsten thread in the seal area and pulling it out after the packages had been sealed and retorted*. The rest of the packages were intact control packages.

Test 3

In this test microholes were made by using XeCl-eximer laser, wavelength 308 nm*. The holes were made in the 220 m thick lid material and had a slightly conical shape. The packages were filled with spaghetti in meat sauce. The packages were filled and sealed with the prepared lid material before retorting.

*) The preparation of the packages has been carried out at the VTT Biotechnology and Food Research Laboratory in Finland.

5 RESULTS

Test 1

Table 1 *Number of Leaking HDPE-Trays Detected by Microhole Tester and Dye-test. The Total Number of Empty Packages were 102 and the Total Number of Packages Filled with Steak were 40.*

	Hole diameter [m]	Number detected by Microhole Tester	Number detected by Dye-test
Empty	12 - 428	102	100
Steak	53 - 307	39	38

The package not detected by the Microhole Tester, was not detected by dye-test either. The hole was probably blocked by food.

Test 2

Table 2 *Number of Leaking PP/EVOH/PP-Trays Detected by the Microhole Tester and Dye-Test. The Total Number of Packages Prepared with Microholes were 104 and the Total Number of Control Packages were 104.*

Package type	Hole diameter [m]	Number detected by Microhole Tester	Number detected by Dye-test
Hole made with tungsten thread	7 - 356	104	98
Control Packages	0 - 256	27	3

27/104 so called intact control packages were also detected as leaking by Microhole Tester. When the tungsten thread is drawn out after retorting the holes are opened, but the holes in the control packages were submitted to food during the retort processing which made them easily blocked by food.

Test 3

Table 3 *Number of Leaking PP/EVOH/PP-Trays Detected by the Microhole Tester.*

Hole diameter [m] (microscoped)	Total number of packages	Number detected by Microhole Tester
13	84	20
18	91	25
21	85	62
25	36	33

The holes in the so called intact packages in Test 2 were submitted to food during retorting. So were the holes in test 3. However, the holes in Test 2 were detected easier than the holes in Test 3. The reason for this must be the difference in food. The holes in Test 3 were also checked in a microscope, to see whether the holes were blocked or not. At smaller holesizes blocking of the hole seems to affect if the holes are going to be detected or not. However, as also larger hole sizes with no blocked holes are undetected, it must mean that another phenomena is also affecting. This is probably because the food is covering the surface in the package, which affects the surface activity.

References

1   L. Axelson, S. Cavlin, and J. Nordström, *Packaging Technology and Science*,   1990,
    3 141-162.

2. L. Axelson, and S. Cavlin, *Packaging Technology and Science*, 1991, **4**, 9-20.

New Packaging Materials

Edible and Biodegradable Food Packaging

S. Guilbert[1] and N. Gontard[2]

[1]CIRAD-SAR FOOD ENGINEERING AND TECHNOLOGY RESEARCH UNIT,
BP 5035, 34090 MONTPELLIER CEDEX 1, FRANCE
[2]ENSIA-SIARC, FOOD ENGINEERING AND TECHNOLOGY RESEARCH UNIT,
BP 5035, 34090 MONTPELLIER CEDEX 1, FRANCE

1 INTRODUCTION

Three main different ways to make "bio-packaging" by using agricultural raw materials are proposed[1] (Figure 1): synthetic polymer/biopolymer mixtures, microbial polymers produced by fermentation of agricultural substrates and finally agricultural polymers used directly as basic packaging material.

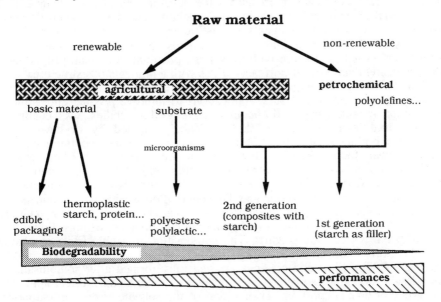

Figure 1 *Different approachs to make "bio-packagings" by using agricultural raw materials (adapted from Gontard and Guilbert[1])*

The first commercial "biodegradable" plastics were developed using a technique involving extrusion mixing of granular native starch (5-20%) and prooxidative and auto-oxidative additives with the synthetic polymer. This technique has been marketed by several firms. Biodegradability of these materials is highly controversial [2,3], their behavior is now classified as "biofragmentation", *i.e.* fragmentation into small molecules. A finest molecular mixture of synthetic polymers and starch-based polymers can be made with gelatinized starch (up to 40-75%), hydrophobic synthetic polymers (polyethylene *etc...*) and hydrophilic co-polymers. The latter compounds act as compatibility agents providing an interface between the starch and the synthetic polymer. Full biodegradability of these materials, as claimed by the manufacturers, is still a topic of discussion.

Microbial polymers such as poly hydroxy butyrate or valerate and polylactic or polyglycolic acids are excreted or stored by microorganisms cultivated on starch hydrolysates or lipidic mediums. Isolation and purification costs are very high for these products that are obtained from complex mixtures and their applications are actually limited by the prohibitively high price. They are recyclable and biodegradable since linkages within polymer chains are formed by natural enzymatic systems which are potential targets of chemical or enzymatic hydrolysis.

Films composed of polymers of agricultural origin (in a natural state or fractionated, *e.g.* whole grains, flours, proteins, starch, fractions...) are economical due to the low cost of raw materials. They are completely biodegradable, and edible when no non food-grade additives are used. "All-starch" packaging made from thermoplastic starch is the most developed among these biopolymers. Standard techniques used for forming synthetic polymer films are generally used (extrusion, injection moulding...). This type of material is very sensitive to water and applications are limited to some short shelf life food packaging, edible packaging and agricultural mulching. The improvement of the water resistance and barrier properties is of first importance for the development of these hydrocolloids based materials. Chemical modification of the polymer and development of specific additives such as crosslinking agents or plastifiers adapted to the polymer structure are now studied[4]. Regarding these developments, proteins which have a "non monotonous" complex stucture with very large potential functional properties are more promising than most of polysaccharides with a homogenous structure.

Edible films or coatings have long been used empirically to protect food products. A few examples of such applications to improve product appearance or conservation include sugar and chocolate coatings for candies, wax coating for fruits and traditional lipoprotein "skins" ("Yuba" obtained by drying the skin formed after boiling soya milk). Solid lipids and oils are also commonly used to cover or coat foods. Edible films are an interesting and often essential complementary parameter to control the quality and stability of many foods. Formulation technology and application of edible films have been reviewed by Guilbert and Biquet[5] and Kester and Fennema[6]. Polysaccharide, protein or lipid materials in various forms (simple or composite material, single-layer or multilayer film) has been proposed for the formulation of edible films or coatings. Edible films are part of the whole food product. The composition of edible films or coatings must therefore conform to the regulations that apply to the food product concerned. Edible films and coatings are generally formed after solubilisation or dispersion in food grade solvent (water, ethanol, organic acid...), direct application on the food or on a support and evaporation of the solvents. Free, self-supporting film can be obtained by standard techniques such as extrusion.

General mechanisms for bio-packaging formation are schematised on Figure 2.

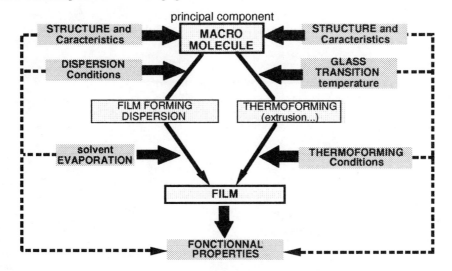

Figure 2 *General mechanism of bio-packaging (film) formation*

2 PROPERTIES AND APPLICATIONS OF "BIO-PACKAGING"

Edible and biodegradable films must cope with a number of specific functional requirements (moisture barrier, solute and/or gas barrier; water or lipid solubility; colour and appearance; mechanical and rheological characteristics; non-toxicity, *etc.*). These properties are dependent on the type of material used, its formation and application. Plasticizers, cross-linking agents, anti-microbial, anti-oxygen agents, texture agents, *etc.* can be added to enhance the functional properties of the film.

The mechanical properties of edible or biodegradable films depend on the type of film-forming material and especially on its structural cohesion. Cohesion depends on the structure of the polymer and especially its molecular length, geometry, molecular weight distribution and the type and position of its lateral groups. The mechanical properties are also linked with the film-forming conditions (type of process and process parameters). For example the puncture strength of gluten based films obtained from a dispersion in ethanol, water and acetic acid is also strongly dependent on film-forming conditions *i.e.* gluten concentration and pH of the film-forming solution[7], as illustrated in Figure 3.

The mechanical properties of amorphous materials are seriously modified when temperatures of these compounds rise above the glass transition temperature (Tg). The glass transition phenomenon separates materials into two domains according to clear structural and property differences, thus dictating their potential applications. Below Tg the material is rigid, and above it becomes viscoelastic or even liquid. Indeed, below this critical threshold only weak, non-cooperative local vibration and rotation movements are possible. Film relaxation relative to temperature follows an Arrhenius time course. Above the Tg threshold, strong, cooperative movement of whole molecules and polymer segments can be observed. These are cooperative structural rearrangement movements. In the Tg (Tg + 100°C) temperature range, these movements are given by the Williams, Landel and Ferry equation [8].

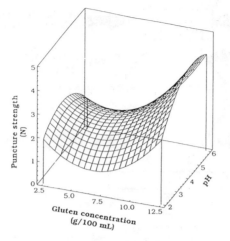

Figure 3 *Effect of pH and gluten concentration of the film forming solution on the puncture strength of a gluten film (adapted from Gontard et al.[7])*

For example this glass transition phenomenon has been demonstrated in gluten films by differential scanning calorimetry and confirmed through dynamical thermomechanical analysis by Gontard et al.[9]. The phenomenon seems to be a crucial physico-chemical parameter for understanding and predicting the behaviour of films such as that formed with gluten.

The mechanical properties of films can be enhanced by plasticization of the polymeric network. There are two different plastifying effects. Internal plasticization is obtained by modifying the chemical structure of the polymer, *e.g.* by co-polymerisation, selective hydrogenation or transesterification when edible lipid or derivative materials are used. External plasticization is obtained by adding agents which modify the organisation and energy involved in the tridimensional structure of film-forming polymers[10]. Reduction of the intermolecular forces between polymer chains facilitates extensibility of the film (less brittle, more pliable) and reduces its Tg. However, this also results in reducing the gas, vapor and solute barrier properties of the film. The illustration of the effect of water on Tg, puncture strength and water barrier property observed for gluten based films are given respectively in Figures 4, 5 and 6 .

Water is the most common plastifier[9,11,12] and is very difficult to control in biopolymers which are generally more or less hydrophilic. Plasticization of biopolymeric films is thus dependent on the usage conditions, especially relative humidity (of environment and packaged products). In isothermic conditions, the addition of plastifiers such as water has theoretically the same effect as increased temperature on molecular mobility. This phenomenon could be utilized to reduce the glass transition temperature of biopolymers to below the decomposition temperature threshold. Standard techniques for synthetic polymer films such as extrusion or injection moulding, could thus be used. However, the drawback of this phenomenon is that it makes biopolymer packaging moisture-sensitive. Their mechanical properties are generally greatly modified by high temperature and/or moisture (ambient or from the packaged product)[11,13].

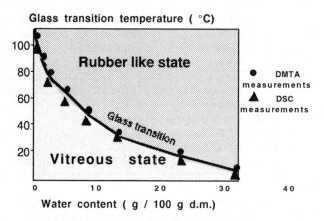

Figure 4 *Glass transition of a gluten film as a function of film water content (adapted from Gontard et al [9])*

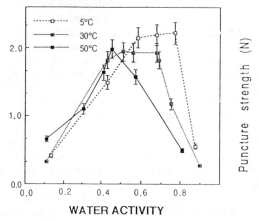

Figure 5 *Puncture strength of a gluten film as a function of water content at 5, 30, and 50°C (adapted from Gontard et al. [11])*

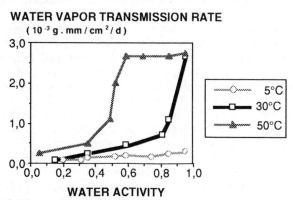

Figure 6 *Water vapour transmission rate of a gluten film as a function of water activity at 5, 30, and 50°C (adapted from Gontard et al.[9])*

Permeability (P) of a permeant (water gas or solute) is only a general feature of films or coatings when the diffusivity (D) of the permeant, the solubility (S) of the permeant molecules are not influenced by permeant content, thus when Fick's and Henry's laws apply. In practice, for most biopolymers the permeant interacts with the film and D and S are dependent on the differential partial pressure. For instance, concerning the water vapor permeability of hydrophilic polymer films, the water solubility and diffusion coefficients increase when the water vapor differential partial pressure increases because of the moisture affinity of the film (nonlinear sorption isotherm) and increased plasticization of the film due to water absorption (Figure 6)· The film thickness can also influence permeability when using film-forming materials that do not behave ideally.

Hence, it should be noted that the permeability of biopolymer films is a property of the film-permeant complex under defined ambient conditions (temperature and humidity).

Using sorption curves, it is quite easy to determine the effect of temperature and relative humidity on solubility of the permeant in the material. In contrast, it is more difficult to determine the nature of the functional relationship between diffusivity and temperature or water content. Various theories, including the free volume theory, have been put forth to explain this functional relationship. According to the free volume theory, molecular diffusion results from redistribution of the free volume in the material. This diffusion is only valid if the size of the free volume faults is greater than the critical value defined by the size of the diffused molecule. This value can be reached after a temperature increase. At T<Tg: mobility is controlled by the preexistence of pores in the glassy material, at T>Tg: polymer relaxation and porosity influence mobility. The free volume theory allows one to qualitatively predict variations of D relative to the difference between the temperature and the glass transition temperature (Tg). This theory is especially valid in the Tg to Tg+100°C range.

Water vapor permeability of some biopolymer-based and synthetic films are given in Table 1. Permeability is clearly high in films formed from hydrophilic materials. These films can only be used as protective barrier layers to limit moisture exchange for short-term applications or in low-moisture foods.

Lipidic compounds are often used to make moisture barrier films and coatings (Table 1). Water is not very soluble or mobile in lipid-based films because of the low polarity and dense, well-structured molecular matrixes that can be formed by these compounds.

Moisture resistance of lipid films is generally inversely related to polarity of the lipids. The moisture barrier capacities of different films can be classified in increasing order of efficiency, as follow : liquid oils < solid fats < waxes[14,15].

There is usually a difference between water vapor and gas permeability (CO_2, O_2) of the same film. According to Banker [10] and Kester and Fennema [16], gas diffusion is crucial for gas permeability, whereas both sorption and diffusion are essential for moisture transfer. Materials with suitable oxygen barrier properties are required to protect oxidizable foods (to reduce rancidity and vitamin loss), but some permeability to oxygen, and especially to CO_2, is essential for fresh fruit and vegetable coatings. Some biopolymer-based packaging has impressive gas barrier properties when they are not moist, especially against oxygen. The oxygen permeabilities of some edible and non-edible films are given in Table 2.

When moisture is present, the macromolecule chains become more mobile which leads to a substantial increase in oxygen permeability· Lipids, which are very often used to delay water transfer also have significant oxygen barrier properties. An increase in the degree of unsaturation or branching and reduction in the length of the lipidic carbon chain lowers oxygen permeability. The following barrier efficiency order was observed[17]: stearic alcohol > tristearine > beeswax > acetylated monoglycerides > stearic acid > alkanes.

The development of packaging or of edible films and coatings with selective gas permeability could be very promising for controlling respiratory exchange and improving the conservation of fresh or minimally processed fruits and vegetables[18]. For example hydrocolloid based films such as gluten films[18] appear to have suitable oxygen barrier properties while remaining relatively permeable to CO_2 (Table 2).

Concerning oxygen and carbon dioxide, the relative solubility of CO_2 in water explains its high permeability in hydrophilic materials. The selectivity coefficient relative to these two gases is therefore dependent on the moisture content of the film.

Table 1 *Water Vapor Permeability of Various Films*

Film conditions	Water Vapor Permeability $(x10^{12}mol.m.m^{-2}.s^1 \cdot Pa^{-1})$	T (°C)	Thickness $(x10^3 m)$	RH %
Starch	142	38	1.190	100-30
Casein - gelatin	34.3	30	0.250	60-22
Wheat gluten and glycerol	5.08	30	0.050	100-00
Wheat gluten and oleic acid	4.15	30	0.050	100-00
Wheat gluten and carnauba wax	3.91	30	0.050	100-00
LDPE	0.0482	38	0.025	95-00
Wheat gluten-beeswax-bilayer	0.0230	30	0.090	100-00
Beeswax	0.0122	25	0.120	87-00
Aluminium foil	0.000289	38	0.025	95-00

LDPE is low density polyethylene

Table 2 *Oxygen and Carbon Dioxide Permeabilities of Various Films*

Film	O_2 Permeability $(x10^{18}mol.m.m^{-2}.s^{-1}.Pa^{-1})$	CO_2 Permeability $(x10^{18}mol.m.m^{-2}.s^{-1}.Pa^{-1})$	T (°C)	Aw
LDPE	1003	4220	23	0.0
HDPE	285	972	23	0.0
Polyamide	12	19	23	0.0
Cellophane	8		23	0.50
Wheat Gluten	1		25	0.0
Pectin	1340	21300	25	0.96
Wheat Gluten	1290	36700	25	0.95
Chitosan	472	8010	25	0.93

HDPE is high density polyethylene
LDPE is low density polyethylene

3 MODIFICATION OF SURFACE CONDITIONS WITH "EDIBLE ACTIVE COATINGS"

Edible active films and coatings could be applied on foods to modify and control surface conditions. The improvement of food microbial stability can be obtained by using edible active layers which have specific antimicrobial properties or pH lowering properties. They can also be used as surface retention agents to limit food additives diffusion in food core.

Edible films and coatings can be used in combination with treatments such as refrigeration and controlled atmosphere to improve the microbiological quality of certain foods[19,20]. For example, calcium alginate or chitosan based films were tested to limit microorganism contamination on the surface of various foods[21,22,23]. These films were found to have a significant effect on microbial growth. For example chitosan based films applied on a model food (Aw=0.97, T=30°C) can delay the development of *Penicilium notatum* for more than seven days after the contamination[24]. The specific antimicrobial activity of calcium alginate coatings has not yet been explained, but could be partially due to the presence of calcium chloride[25],while the effect of chitosan can be attributed to the fact that chitosan is a cationic molecule[26].

The improvement of food microbial stability can also be obtained by reducing surface pH. This can be achieved by using films or coatings that immobilize either specific acids or charged macromolecules [27].

As a matter of fact, edible films and coatings can be used as food preservative media (particularly for antioxygen and antifungic agents) and as surface retention agents to limit preservative diffusion in the food core[6,20,28-33]. Maintaining a local high effective concentration of preservative to reduce aerobic contamination and/or oxygen influence may allow, to a considerable extent, a reduction of its total amount in the food for the same effect. It is then important to be able to predict and control surface preservative migration. Preservative diffusion through edible films is influenced by various parameters: film characteristics (type, manufacturing procedure), food characteristics (pH, a_w), storage conditions (temperature, duration, *etc.*) and solute characteristics (hydrophilic properties, molar mass).

The retention of α tocopherol in gelatin films used on the surface of margarine was evaluated[29].After 50 days storage, no migration was observed when the film was pretreated with a cross-linking agent (tannic acid), whereas without a film, a tocopherol diffusion was found to be as high as 10-30 10^{-11} $m^2.s^{-1}$.

Guilbert[29] also investigated the retention of sorbic acid in gelatin and casein films treated respectively with tannic or lactic acid, and placed over an aqueous model food system (a_w = 0.95). After 35 days at 25°C, a retention of 30 % was observed with the treated casein film.

Sorbic acid retention has been studied in zein-based[30], casein-based[28,29], gluten-based and pectin-based films[34] and in composite polyoside-lipid derivative films[33]. The sorbic acid permeability values determined for these films can also be compared with published apparent diffusion values for sorbic acid in food systems. In an intermediate moisture agar model system, Guilbert *et al*[35] reported a value of 2.0 x 10^{-10} m^2/sec. Torres *et al* [30] found with an intermediate moisture cheese analog a value of 1.0 x 10^{-10} m^2/sec. The sorbic acid diffusivity in edible films were found to be 150- to 300-fold lower than those determined for model intermediate moisture foods. These few examples indicate that edible films could be used for additive retention on the surface of food products.

Microbiological analyses generally confirm the efficiency of preservative retention within surface coating. Zein-based edible films containing sorbic acid double the shelf-life of intermediate moisture cheese analogs before the appearance of microorganisms[27, 30]. Significant improvement of the microbial stability of intermediate moisture fruit coated or not with edible coating containing sorbic acid was observed[29]. The coatings efficiency were in the following order : carnauba wax + sorbic acid > carnauba wax > casein + sorbic acid > casein > no coating. New results[24] show that pectin films, gluten films and gluten/monoglyceride derivative composite film containing sorbic acid can delay the development of *Penicilium notatum* on a model food (Aw=0.97, T=30°C) respectively for more than one, two or four days after the contamination as compared to a standard where the sorbic acid is only deposited at the food surface.

4 CONCLUSION

The studies presented here have demonstrated a number of characteristics of food macromolecules that make them suitable for the formation of different types of wrappings and films. The use of these properties and their ability to be modified and controlled opens a new field of application for these macromolecules in a non-food sector, for the manufacture of biodegradable packaging. Research and development is still required to develop packaging material composed entirely of renewable biodegradable macromolecules from agricultural products that have good performance characteristics and are economical.

Edible superficial layers provide a supplementary and sometimes essential means to control physiological, microbiological and physicochemical changes in food products which could be called "active edible layers concept". This by controlling gas exchange (water vapor, oxygen, carbon dioxide, *etc.*) between the food product and the ambient atmosphere, or between components in a heterogeneous food product, and by modifying and controlling food surface conditions (pH, level of specific functional agents, *etc.*) . It should be stressed that the characteristics of the film or coating and the application technique must be adapted to each specific utilization.

References

1. N. Gontard and S. Guilbert, 'Food Packaging and Preservation', M. Mathlouthi ed, Blackie Professional, 1994, p 159.
2. M. Vert, *Caoutchoucs et plastiques*, 1991, **706**, 71.
3. L.R. Krupp and W.J. Jewell, *Environ. Sci. Technol.,* 1992, **26** (1), 193.
4. S. Guilbert and J. Graille, Acta of '*1er colloque national sur les valorisations non alimentaires des grandes productions agricoles*', 1994, 18-19 Mai, Nantes.
5. S. Guilbert and B. Biquet, 'L'emballage des denrées alimentaires de grande consommation', J.L. Multon and G. Bureau eds, Lavoisier, Paris, 1989, p. 320.
6. J.J. Kester and O. Fennema, *Food Technol.* ,1986, **12**, 47.
7. N. Gontard, S. Guilbert and J.L. Cuq, *J. Food Sci.,* 1992, **57** (1), 190.
8. M.L. Williams, R.F. Landel and J.D. Ferry, *J. Amer. Chem. Soc.,* 1955, **77**, 3701.
9. N. Gontard, S. Ring, S. Guilbert and L. Botham, *J. Food Sci.,* submitted.
10. G.S. Banker, *J. Pharm. Sci.* 1966, **55**, 81.

11. N. Gontard, S. Guilbert and J.L. Cuq, *J. Food Sci.*, 1993, **58**,(1), 206.
12. D. Rasmussen and B. Luyet, 'The Technology of Plasticizers' J. Wiley Inters. New-York, 1969.
13. R.L. Evangelista, Z.L. Nikolov, W. Sung, J. Jane and R.J. Gelina, *Ind. Eng. Chemic. Res.*,1991, **30**, 1841.
14. N. Gontard, C. Duchez, J.L. Cuq and S. Guilbert, *Int. J. Food Sci. Technol.*,. 1994, **29**, 39.
15. S.L. Kamper and O. Fennema, *J. Food Sci.*, 1984, **49** (6), 1478.
16. J.J. Kester and O. Fennema, *J. Amer. Oil Chem. Soc.*, 1989, **66** (8) 1147
17. J.J. Kester and O. Fennema, *J. Amer. Oil Chem. Soc.*, 1989, **66** (8), 1129.
18. N. Gontard, R. Thibaut and S. Guilbert, *J. Food Sci.*, 1994, submitted
19. S. Guilbert., 'Food Preservation by Combined Processes', 1994. Report FLAIR Concerted Action n°7, p 65.
20. S. Guilbert, N. Gontard and A.L. Raoult-Wack, 'Food Preservation by Moisture Control', J. Welty ed, Technomic Publishing Co, USA, 1994.
21. S.K. Williams, J.L. Oblinger and R.L. West., *J. Food Sci.*, 1978, **43**, 292.
22. A. El Ghaouth, R. Ponnampalam and J. Arul, *87th Annual Meeting of the American Horticultural Society,* Tucson, Arizona, 1990, **25**, 1086.
23. A. El Ghaouth, R. Ponnampalam and J. Arul, *J. Food Proc. Pres.*, 1991, **15**, 359.
24. S. Guilbert, non published results.
25. C.R. Allan and L.A. Hadwiger, *Exp. Mycol.*, 1979, **3**, 285.
26. P. Stössel and J.L. Leuba, *Phytopath. Z.*, 1984, **111**, 82.
27. J.A. Torres and M. Karel, *J. Food Proc. Pres.*, 1985, **9**, 107.
28. S. Guilbert, 'Food Packaging and Preservation', M. Mathlouti ed, Elsevier Applied Science Publishers, New-York, 1986, p. 371.
29. S. Guilbert, 'Food Preservation by Moisture Control', G. Seow ed, Elsevier Applied Science Publishers, New-York, 1988, p. 199.
30. J.A. Torres, M. Motocki and M. Karel, *J. Food Proc. Pres.*, 1985, **9**, 75.
31. F. Vojdani and J.A. Torres, *J. Food Proc. Pres.*, 1989, **13**, 417.
32. F. Vojdani and J.A. Torres, *J. Food Proc. Eng.*, 1989, **12**, 33.
33. F. Vojdani and J.A. Torres, *J. Food Sci.*, 1990, **55**, 841.
34. F. De Savoye, F. Dalle Ore, N. Gontard and S. Guilbert, International Colloque "Le froid et la qualité des légumes frais" 1994, 7-9 September, Brest, France.
35. S. Guilbert, A. Giannakopoulos and J.C. Cheftel, 'Properties of Water in Foods in Relation to Quality and Stability', D. Simatos and J.L. Multon eds., Martinus Nijhoff Publishers, Dordrech, Netherlands, 1985, p. 343.

Aroma Compounds and Water Vapour Permeability of Edible Films and Polymeric Packagings

F. Debeaufort, N. Tesson, and A. Voilley

ENSBANA, LABORATOIRE DE GÉNIE DES PROCÉDÉS ALIMENTAIRES ET BIOTECHNOLOGIQUES, UNIVERSITÉ DE BOURGOGNE, 1, ESPLANADE ERASME, 21000 DIJON, FRANCE

1. ABSTRACT

The permeabilities of water vapour and aroma compound through low density polyethylene, cellophane and edible films were measured using a dynamic method coupled with detection by gas chromatography. Water vapour and 1-octen-3-ol transfer rates increased the amount of penetrant sorbed within the film, particularly for hydrophilic films such as cellophane or edible films. Moreover aroma sorption in hydrophilic polymers strongly depended on the water activity of the atmosphere. Neither diffusion coefficient nor amount sorbed, was sufficient to explain the differences observed between edible film permeabilities.

2. INTRODUCTION

One of the quality criteria of foodstuff is aroma; meaning volatile compounds with low molecular weight. To prevent flavour changes, aroma transfers could be reduced by the use of appropriate packaging[1]. Several phenomena could occur during the storage of foodstuffs such as :

    - "Scalping", i.e the fixation of volatile compounds on the surface of film packaging

    - Migration of compounds through films due to adsorption at the inner surface, diffusion within the polymer and desorption at the outer surface. The knowledge of solubility and diffusion coefficients of volatile compounds in packaging materials allows one to predict and limit the loss of aroma intensity during storage.

    Most works deal with the influence of operating conditions (aroma concentration, nature of the medium, temperature) and the nature of the volatile on sorption and diffusion in polymers[2,3]. The effects of nature, structure and crystallinity of polymers were largely studied[4,5,6]. To reinforce the barrier efficiency of foodstuff packaging, edible films and coatings directly applied on the surface of food can be used. Until now, most works on edible packagings deal with the water vapour transfers.

    The objective of this work was the study of transfer and sorption of the water vapour and an aroma compound in edible films compared to usual packaging.

3. MATERIAL AND METHODS

3.1 Material

The volatile selected was 1-octen-3-ol (Fluka) whose flavour is typical of mushroom, molecular weight, 128. The hydrophobicity constant, log P, is 2.6 and was calculated from the Rekker method[7].

A cellophane 300 P (Courtauld's films, 20 µm) and a low density polyethylene FF30 (EniChem Polymers France, 25 µm, 37 % crystallinity) were used as hydrophilic and hydrophobic polymer films respectively.

Edible films were obtained after solubilization and homogenisation of 4.25 g of methylcellulose (MC-Benecel MO21, Aqualon France) or vital gluten (Manito, Eurogerm) in a water-ethanol (3:1) mixture at 75 °C under magnetic stirring. 1.4 g of polyethylene glycol 400 (Merck), used as a plasticiser, was added to the flux forming solution before spreading out on glass plates. Films were dried 12 hours at room temperature and under relative humidity before unsticking. Film thickness was between 20 and 30 µm.

Before any measurements, films were stored one week over P_2O_5, except for gluten films at 53 % RH.

3.2 Methods

1-Octen-3-ol flux (Foct) and water vapour transfer rates (WVTR) were determined with a dynamic method. Films were placed in a permeation cell where the inner side of the film was swept with a carrier gas containing known vapour concentration of water and aroma. The technique and apparatus used were described in a previous work[8].

Moisture sorption isotherms were determined at 25°C from the "microatmosphere method"[9] at nine activities (0.11 ; 0.22 ; 0.33 ; 0.44 ; 0.53 ; 0.57 ; 0.71 ; 0.84 and 0.91) using saturated salt solutions. The method determines the water content of films at the equilibrium.

1-Octen-3-ol sorption was obtained from the derived method previously described : 0.2 g of films were exposed to the atmospheres with a constant aroma concentration and different relative humidities. The total of volatiles sorbed was determined by weighing and the amount of 1-octen-3-ol sorbed was measured by gas/liquid chromatography after hexane extraction. A Chromapack CP 9000 gas chromatograph was equipped with two detectors mounted in series, respectively a Thermal Conductibility Detector (TCD) and Flame Ionisation Detector (FID). The packed column was a Carbowax 20 M W-AW 100-120 mesh. Temperatures were 130°C isothermal in oven, 200°C for detector and 190°C for injection. Gas flow rates were 30 ml/min for helium, 26 ml/min for hydrogen and 260 ml/min for air.

WVTR and 1-octen-3-ol flux were defined as the relation of the weight of permeated vapours to the product of area exposed and time, and were expressed in g m^{-2} h^{-1}. Amount sorbed within the polymer was expressed as mg of volatile per gram of dry matter (mg/g dm). For each determination, at least three replicates were made, and data means were tested for statistical significance at the $p < 0.05$ level using the Student-Newman-Keuls test.

4. RESULTS AND DISCUSSION

Tables 1 and 2 present the results of WVTR and 1-octen-3-ol flux at 25°C for vapour concentration differentials equals 4.5-20.8 mg of water vapour/ml He and 0-1.5 µg of aroma/ml He respectively. The amount of volatile sorbed (Qw) was determined by weighing at the steady state of transfer.

Table 1 *WVTR as a Function of Amount Sorbed (Qw) and Film Nature.*

Hydrophobicity	Film	Qw (mg/ g dm)	WVTR (g m^{-2} h^{-1})
+			
	LDPE	< 0,1	< 0,11
	Gluten	60	10,91
	Methylcellulose	60	19,30
−	Cellophane	141	26,21

Hydrophobicity of films was determined from hexane sorption and contact angle measurements between a water droplet and flux surface. WVTR seems to follow the hydrophobicity of the polymer and water content (Qw). Water vapour transfer strongly depends on the sorption and diffusion as - described by the model of sorption - diffusion is commonly used to define the permeability.

Moisture sorption isotherms of edible films containing 25 % of PEG 400 to dry matter do not differ (Figure 1). Moreover, no hysteresis was observed between sorption and desorption of water in methylcellulose and gluten films. However, when edible film water contents at steady state of transfer are the same, WVTR of gluten films is two times lower than those of methylcellulose. This difference can be explained either by diffusion coefficient of water within the polymer or by structure. A previous work showed that water concentration profile within edible films is not linear and diffusion is concentration dependent due to the formation of clusters[10]. Moreover, some works displayed that methylcellulose films are in a rubbery state at 25°C[11] though gluten films are in a glassy state[12] for the same concentration of water and plasticizers. It is well known that diffusion strongly increases above the Tg, thus in rubbery state, which could explain the differences observed between MC and gluten WVTR's but water vapour affects the sorption and transfer of apolar compounds such as volatile aromas. Indeed in presence of water and for a constant aroma concentration differential, the 1-octen-3-ol flux exponentially increases with the WVTR[8].

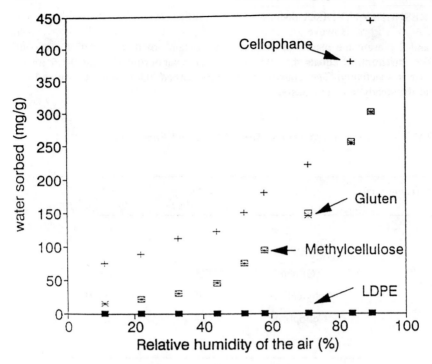

Figure 1 *Moisture sorption isotherms of edible and polymeric films at 25°*

Table 2 : *1-Octen-3-ol Flux (Foct), Amount Sorbed (Qoct) and Diffusion (Doct) through Edible and Packaging Films at the Steady State of Transfer.*

Film	Foct $g\ m^{-2}\ h^{-1}$	Q oct $mg/g\ dm$	Doct $10^{-14}\ m^2\ s^{-1}$
Methylcellulose	0.7884	100	1.41
LDPE	0.3290	4.6	n.d
Gluten	0.0299	58	5.75
Cellophane	< 0.0002	1.2	n.d

Contrary to WVTR, Foct does not follow neither hydrophobicity of films, nor amount of aroma sorbed. However, for hydrophilic films, Foct is exponentially proportional to the quantity of aroma sorbed within the polymer at the steady state of transfer. The methylcellulose film is 25 times more permeable than gluten although Qoct is only 1.7 time larger. The apparent diffusion coefficient was determined from the permeation kinetic and the following equation (1) applied for permeation[13].

$$ Doct = \frac{1^2}{7.199\ t^{1/2}} \tag{1} $$

Where l is the film thickness and $t^{1/2}$ the time required to the permeation rate F to reach 0.5 F∞ which is the permeation rate at the steady state of transfer. From Table 2, diffusion follows the opposite direction of 1-octen-3-ol flux and sorption, and therefore cannot explain differences observed between edible films. As supposed for water transfer, the structure of polymer seems to be preponderant for the aroma transfer and sorption. Moreover, the relative humidity of the atmosphere affects the 1-octen-3-ol sorption in hydrophilic films, but not in low density polyethylene which is apolar (Figure 2).

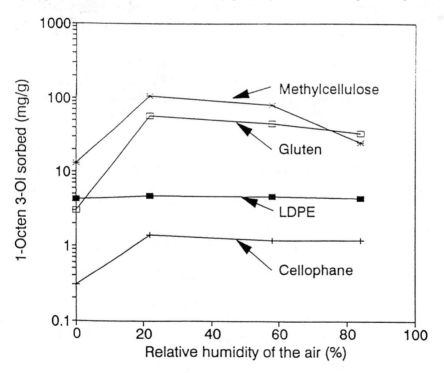

Figure 2 *1-octen-3-ol sorption in edible and polymeric films as a function of the relative humidity of the air at 25° (Aroma concentration = 5,7 µg/ml air)*

On the contrary of moisture sorption, for low relative humidity, methylcellulose sorbed twice as much aroma as gluten. For all hydrophilic films, aroma sorption is very low at relative humidity close to zero. This is probably due to the absence of water. Indeed, water involves a plasticization of the polymer network inducing a polymer inflating and making sorption easier. Above 22 % RH, the more RH increases, (i.e. the water content and hydrophilicity of cellophane and edible film), the lower is the amount of aroma sorbed. This competitivity of sorption can be explained by the stronger affinity of the polymer for water than the apolar compound. Thus, if the 1-octen-3-ol flux through edible films increases with the WVTR and the water content of film, this is not caused by the sorption or the diffusion of the aroma compound, but by a large plasticization phenomena and by a sweeping of 1-octen-3-ol with water vapour.

5. CONCLUSIONS

The water vapour transfer rate through either hydrophilic packagings such as cellophane and edibles films, or hydrophobic wrappings mainly depends on the hydrophobicity of the polymer. For edible flux, sorption itself cannot explain differences observed, but network structure and glass transition can. For aroma compounds, transfer does not follow neither the amount sorbed, nor the diffusion coefficient, nor the hydrophobicity of the polymer. Moreover water strongly increases 1-octen-3-ol flux though it decreases sorption. Thus the sorption-diffusion model-itself cannot describe the aroma and/or water vapour permeability of packagings, particularly of edible films.

Acknowledgements
        The authors wish to thank Isabelle Winter and David Rigaud for their collaboration in this work, and Pechiney Emballage Alimentaire, Eurogerm and Anvar Bourgogne for their technical and financial support.

References

1. S.M. Mahoney, R.J. Hernandez, J.R. Giacin, B.R. Harte, J. Miltz, *J. Food. Sci.*, 1988, **53**(1), 253.
2. T.J. Nielsen, M. Jägerstad, R.E. Öste, *J. Sci. Food. Agric.*, 1992, **60**, 377.
3. G.D. Saddler, R.J. Braddock, *J. Food. Sci.*, 1991, **55**(6), 35.
4. Z.N. Charara, J.W. Williams, R.H. Schmidt, M.R. Marshall, *J. Food. Sci.*, **57**(4), 199.
5. J.B. Konczal, B.R. Harte, P. Hoojjat, J.R. Giacin, *J. Food. Sci.*, 1992, **57**(4), 967.
6. J. Letinski, G.W. Halek, *J. Food. Sci.*, 1992, **57**(2), 481.
7. R. Rekker, "Pharmaco Chemistry Library 1", Elsevier Scientific Publishing Co., London, 1977.
8. F. Debeaufort, A. Voilley, *J. Agric. Food. Chem.* (submitted), 1994a.
9. P. Jowitt, P.J. Wagstaffe, "The certification of the water content of microcrystalline cellulose at ten water activities",Community Bureau of Reference, BCR CRM 302, EUR 12429 EN, 1989.
10. F. Debeaufort, P. Meares, A. Voilley, *J. Mem. Sci.*, (in press), 1994b.
11. C.M. Koelsch, T.P. Labuza, Int. *J. Food. Sci.* Technol., 1992, **25**, 404.
12. N. Gontard, Thèse de doctorat, Université de Montpellier, 1991.
13. P.J. Felder, *J. Mem. Sci.*, 1978, **3**, 15.

Active Packaging

Active Packaging for Food Quality Preservation in Japan

T. Ishitani

JAPAN INTERNATIONAL RESEARCH CENTER FOR AGRICULTURAL SCIENCES, MINISTRY OF AGRICULTURE, FORESTRY AND FISHERIES, TSUKUBA, JAPAN

1. INTRODUCTION

To date, a range of different packaging materials, their subsidiary materials, and packaging technologies have been developed to facilitate quality improvement and preservation and to characterize packaged foods. The technology development of packaging has been made mainly by the activities of private sectors and various kinds of functional materials related to the active packaging have been developed and patented in the latter half of 1980s, especially from 1987 to 1989.

Packaging technology occupies a very important position in the Japanese industry and more than 70 billion US$ of packaging materials are produced and shipped per year. Until now, many basic problems had to be solved and most of them still remain especially in the field of food packaging, and cooperation systems among universities, national institutes, prefectural institutes and private sectors are urgently needed in Japan. Fortunately, 3 years ago the Japanese Society of Packaging Science and Technology started to act as an academic society for packaging study. Now we are expecting various academic works to solve the basic problems through newly established cooperation systems.

2. MAJOR FUNCTIONAL FOOD PACKAGING MATERIALS AND THEIR FUNCTIONS

Until now, many different types of functional materials and their subsidiary materials have been developed (Table 1), and contributed to the food quality improvement, preservation and characterization of packaged foods. Most of them are new materials and have new functions using lamination, coating or incorporating useful materials.

2.1 Gas Barrier Materials

There is a group of functional packaging materials which has exceedingly high barrier properties for oxygen, carbon dioxide, water vapor and volatiles. These gas barrier materials are widely used for the prevention of oxidation, discoloration as well as fungal growth, and for moisture-proof packaging purposes.   Oxygen barrier property is an important target of new packaging materials development, and a series of high barrier packaging materials has been newly developed and introduced to commercial use.

Recently, several materials with high barrier performance have been brought to our attention, such as laminates of SiOx-vaporized and deposited film, co-extruded multi-layer films of EVOH or PVDC, formed containers with EVOH or steel foil having reduced moisture effects. These materials, such as SiOx-deposited films and, co-extruded films and sheets of PP/EVOH/PP, PP/PVDC/PP etc. are being used as retortable and microwavable high barrier packaging materials. It should be noted that SiOx-deposited film reduces the problems of waste disposal.

Table 1. *Major "Value Added" Packaging Materials for Food Quality Preservation*[1]

Barrier properties :
   Oxygen barrier : EVOH. PVDC. OV (BOV). PAN. NMXD6. Al metalized.
             SiOx-deposited
   Carbon dioxide barrier : EVOH. SiOx-deposited
   Water vapor barrier : HDPE. OPP. PVDC. Al-metalized. SiOx-deposited
   Light (UV) barrier : Titanium oxide fine powder-incorporated. Iron oxide powder. PET.
             Hydroxybenzophenone-incorporated, etc.
   Volatiles barrier : PET. PC. PVDC. EVOH. OV (BOV). SiOx-deposited, etc.
   Insulation : Foamed plastics. Al metalized
   Smoke-permeable oxygen barrier : Polymer alloy of PA, etc.
   High gas permeability : LLDPE. OPP. PVC. EVA. PS. Polybutadiene. Polymethyl
             pentene. Micro-perforated films, etc.
Anti-microbes : Silver-zeolite incorporated. Quaternary ammonium salts-incorporated, etc.
Anti-fog : Detergent-incorporated plastics, BOV, etc.
Water absorption : Water-absorbing polymer incorporated sheets. RH adjusting sheets.
             Sheets for semi-drying of food. Desiccant-incorporated film
Gas absorption : Inorganic filler (zeolite etc.)-incorporated. Chemicals-incorporated, etc.
Heat resistance : PVDC. PP with inorganic filler. CPET. PMP. PC. PPO. PPS. PSF. PI, etc.
Plastic odor free : Low plastic odor sealants
Low flavor sorption : PET-G. PAN. Heat sealable EVOH
Gas scavenging or generating
   Volatiles generating : Ethanol vapor. Phenolic antioxidant. Allyl isothiocyanate.
             Hinokitiol. Flavor components, etc.
   Volatiles scavenging : Ethylene (potassium permanganate, zeolite, active charcoal, etc.)
   Oxygen scavenging : Inorganic type (Ferrous type). Organic type (reduction, active
             charcoal, etc.)
   Moisture absorption (desiccant) : Calcium oxide. Silica gel. Calcium chloride, etc.

2.2 Flavor Barrier Packaging Materials :
   In order to maintain food flavor or good special fragrance, it is necessary not only to confine flavor substances with barrier materials, but also to prevent volatile substances from dissolving into the sealing layer. Particularly, a major volatile constituent such as limonene in orange juice, a hydrocarbon terpenoid compound, dissolves easily in large amounts in the innermost sealant layer made of polyolefin, resulting in the lost aroma or

balance changes in the flavor. The best way to solve this sort of problem is to use sealing materials that do not have affinity for hydrocarbon volatiles (Figure 1)[2,3]. For this particular purpose, new sealing materials including polyester (PET-G), polyacrylo-nitrile (PAN) and EVOH have been developed and PET sealant has been used in practice in packaging for fruit juices, alcoholic beverages, milk and toiletries, etc. The mechanism of volatile sorption has been studied intensively by Prof. Osajima's group of Kyushu university[4].

These sealants also contribute in reducing the undesirable odor from plastic materials. Polyolefin sealants have some undesirable odors and tastes, and such odors and tastes from food packaging materials have to be reduced as the demand for high quality foods, especially for microwave cooking, increases.

2.3 Gas Permeable Packaging Materials :

A range of freshness-keeping films to create an environment with low-oxygen and high-CO_2 modified atmosphere, favorable to the maintenance of freshness of fruits and vegetables, skillfully exploits gas permeability of packaging materials. Most of the packaging materials used for this purpose are thin films with relatively high gas permeability such as polyethylene, polypropylene, polystyrene, ethylene vinylacetate copolymer, soft polyvinyl chloride, and polybutadiene.

Figure 2[5] shows the positions of oxygen and water vapor permeabilities of each plastic films and

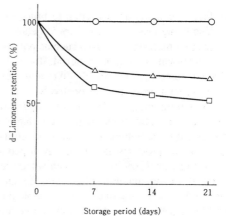

Figure 1. *Changes in d-limonene retention of mandarin orange juice in plastic laminated pouches with PET and PAN sealants during storage at 3°C*

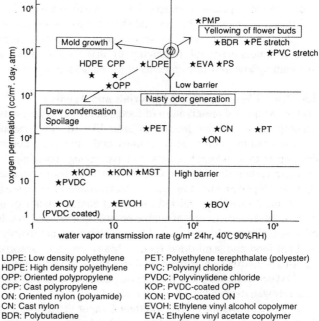

LDPE: Low density polyethylene
HDPE: High density polyethylene
OPP: Oriented polypropylene
CPP: Cast polypropylene
ON: Oriented nylon (polyamide)
CN: Cast nylon
BDR: Polybutadiene
PMP: Polymethyl pentene
BOV: Biaxially oriented vinylon
OV: PVDC-coated oriented vinylon

PET: Polyethylene terephthalate (polyester)
PVC: Polyvinyl chloride
PVDC: Polyvinylidene chloride
KOP: PVDC-coated OPP
KON: PVDC-coated ON
EVOH: Ethylene vinyl alcohol copolymer
EVA: Ethylene vinyl acetate copolymer
PS: Polystyrene
PT: Plain cellophane
MST: Polymer-type moisture-proof cellophane

Figure 2. *Oxygen and water vapor permeability of various plastic films and factors for quality change of broccoli*

the factors causing deterioration to broccoli at 15°C. Films with invisible micro-pinholes are also being used practically for vegetable and flower packaging purposes. Furthermore, many other materials are used such as polymethyl pentene with very high gas permeability, and functional packaging materials for vegetables containing surfactants, gas absorbents and anti-microbial agents to give new functionalities of anti-fog, gas absorption and anti-microbial properties, respectively.

For fermented foods, cheese, yogurt and fermented soybean 'Natto' fermented in packages, or lactic drinks with viable lactic acid bacteria that require oxygen permeation and moisture retention, new package containers with appropriate gas permeability for better quality products have been developed and are in use.

A functional specific film with selective gas permeability is represented by smokable synthetic casing material which was developed by Kureha Chemical Co.Ltd. This film permeates flavor components when smoked at a higher temperature and high humidity, and does not permeate oxygen at low temperatures[6]. This film is made by polymer alloy of polyamide and used for manufacturing of smoked ham, sausages and fish cakes etc.

2.4 Moisture-Humidity Adjusting Materials :

The moisture-absorbing packaging materials utilizing moisture absorbing polymers are widely applied for the purposes of removal of excess moisture, prevention of drying, and moisture adjustment, such as bottom sheet for absorbing drips from meat or fish, cover sheet for preventing dew formation of warm take-out foods like hamburger, fried chicken and pizza etc., packaging materials for preventing weight loss and withering of vegetables and flowers during distribution, and sheet for partial drying of fishes at low temperatures.

Another type of desiccant incorporated in moisture absorbing packaging materials are used for moisture-proof packaging of dried foods and medicines for the purpose of maintaining low moisture contents of the packaged objects.

2.5 Other High-Functional Materials and Agents :

For the quality preservation of foods, many packaging materials and agents that absorb, decompose and remove hazardous gases like ethylene, carbon dioxide, odor of ammonia, sulfur-containing compounds, amines and aldehydes generated from foods have been developed as deodorant agents and packaging materials. And also many agents that generate carbon dioxide gas, ethanol vapor, arylisothiocyanate and hinokitiol vapor have been developed and used for quality preservation of processed foods and fresh produce.

Table 2 shows specialized functional characteristics of oxygen absorber. Many types of oxygen absorbers are available on the market and produced at the level of several million US$ annually in Japan. Oxygen absorbers are mainly used for quality preservation of dried food and semi-dried food. Many oxygen absorbers are widely used for short-term distribution of fresh food like row fish under low temperature.

Oxygen absorbing tray for aseptic packaged cooked rice has been developed and used with nitrogen exchange packaging method.

Figure 3 shows the effects of ethanol vapor generators and water activities of packaged food on the growth of yeast[7]. The combined effect of oxygen absorption and ethanol vapor was observed under higher water activities on several species.

Temperature-time indicators act as kinds of functional agents for monitoring chilled temperatures in the course of distribution and marketing of fresh foods and may assume important technology in the near future.

Table 2. *Specialized Functional Characteristics of Oxygen Absorber*

Absorption speed :   Quick absorption. Intermediate absorption velocity. Slow absorption
Dependence on food moisture : High moisture food (dependent type).
               Intermediate moisture food. Low moisture food (independent type)
Specific function plus : O_2 absorption + CO_2 absorption
               O_2 absorption + CO_2 generation
               O_2 absorption + ethanol vapor generation
Appearance :   Sachet. Very small sachet. Tablet. Sheet (card). Tray.
               With oxygen indicator
Raw materials :   Active charcoal. Reductons. Iron powder. Iron oxide powder, etc.
Purpose :   For frozen food. For chilled food. For microwavable packaged food.

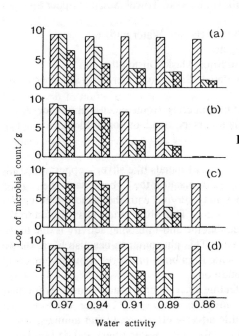

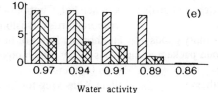

Figure 3. *Comparison of inhibitory effects of two types of ethanol vapor generator on the growth of yeasts*

Microbial counts at the time when microbial count in control at each water activity reached the maximum level were compared.
(a) *Hansenula anomala,* (b) *Saccharomyces cerevisiae,* (c) *Torulaspora delbrueckii,* (d) *Zygosaccharomyces rouxii,* (e) *Candida tropicalis,* ▨ : control, ◩ : ethanol vapor generator, ▧ : O_2 absorption type − ethanol vapor generator

3. THE DEVELOPMENT AND APPLICATION OF ANTI-MICROBIAL PACKAGING MATERIALS

3.1 Introduction

Since 1987, functional packaging materials have become a booming issue in technical development, and new films have been introduced into the market.

The technology of controlling undesirable microorganisms by incorporating or coating anti-microbial substances onto daily goods such as cloths, kitchen utensils and sanitary goods, or packaging materials and industrial materials like water-treatment filters, attracts much attention in Japan (Table 3)[8].

The anti-microbial agents that are applied to those uses require safety considerations, even if the materials containing them do not come in direct contact with human ingestion. For uses involving food packaging or direct human contact, the agents that assure their safety must only be used.

Generally, it is recognized that metallic ions of silver and copper, the quaternary ammonium salts, and natural compounds like hinokitiol are among safe anti-microbial agents. In food-use materials, the agents should not only be safe but also difficult to migrate into foods. Ag-substituted zeolite (Ag-Zeolite) is most commonly used as an anti-microbial agent incorporated into plastic materials. This Ag-Zeolite has been developed in Japan and related subjects are to be discussed.

Table 3. *Applications of Anti-microbial Materials*

Textile and cloths : Bed sheets. White overall. Underwear. Towel. Socks. Slippers.
                         Shoes, etc.
Kitchen utensils :   Kitchen bat. Chopping board. Basket. Water purifier.
                         Scrubbing brush. Dustbin, etc.
Sanitary goods :   Toothbrush. Filter of humidifier. Mask. Dustcloths, etc.
Food packaging materials : Tray (lunch box). Pouch. Paper container. Drip absorber.
                         Wrap film, etc.
Others :   Filters & packing for food machineries. Home electric appliances.
                         Artificial teeth. Mat for plant nursery. Sandbox for children, etc.

3.2 Anti-microbial Activity of Silver Ion.

When bacterial culture is spread over the surface of metals like silver, copper and crude nickel, compared with the surface of paper, glass or plastics, the bacteria die after 3-4 days on the metal surface, whereas they may survive with the same viable counts after 2 weeks on the surface of non-metal materials. This observation is attributed to the action of the metal itself, and it was confirmed that the anti-microbial activity is due to the ions of minute quantity formed from the metals. This phenomenon has also been known from the fact that strap of train or door-knob made from brass or copper show low levels of viable bacterial counts under highly humid conditions.

The copper ion has the activity of disinfecting algae, microbes and viruses, while it is indispensable for life as a constituent of metallic enzymes. Copper does not get concentrated by living beings and thus has little adverse effects on higher animals, which renders the ion as relatively safe among metals. Recently, water filters and daily utensils utilizing anti-microbial activity of copper ion have been commercially available.

Amongst metallic ions, the silver ion has the strongest anti-microbial activity (Table 4), and that of copper is relatively strong. Metallic silver does not release ion easily compared with copper, and its anti-microbial activity is not quite as strong in its metallic state. Silver is a safe and relatively inert metal, and therefore often used in direct human contact as dish and plate, fork and spoon, and artificial teeth etc.

Silver is contained in earth crust at a low concentration below 0.1ppm and used as photographic materials, electric lead wire, kitchen utensils, artificial teeth, silver plating, paints, ornaments, fine art crafts, coins and keys. Also, it is used as an anti-microbial agent in the form of medicine and water-treatment agent. Silver exists as ion when dissolved in water, and becomes insoluble by reacting with halogens quite easily. This is why the distribution of silver is limited in nature. Silver gets strongly absorbed by magnesium oxide, clay minerals and organic metal compounds, and concentrated in aquatic organisms, but to a lesser extent than in the case of mercury. No report is seen with regard to carcinogenicity and mutagenicity of silver. The standard for silver content in drinking water has been set at less than 50 ppb in United States on the basis of a silver-containing medicine which causes argiria symptoms.

Table 4. *Minimum Inhibitory Concentration of Mineral Ions to Salmonella typhi at $37^{\circ}C$*
(Morality)

$Na^{\cdot}$	$K^{\cdot}$	$NH_4^{\cdot}$	$Li^{\cdot}$	$Sr^{\cdot\cdot}$	$Ca^{\cdot\cdot}$	$Mg^{\cdot\cdot}$	$Ba^{\cdot\cdot}$	$Mn^{\cdot\cdot}$	$Zn^{\cdot\cdot}$	$Al^{\cdot\cdot\cdot}$	$Fe^{\cdot\cdot}$
1.0	1.0	1.0	0.5	0.5	0.5	0.25	0.25	0.12	0.001	0.001	0.001

$H^{\cdot}$	$Pb^{\cdot\cdot}$	$Ni^{\cdot\cdot}$	$Co^{\cdot\cdot}$	$Au^{\cdot\cdot}$	$Cd^{\cdot\cdot}$	$Cu^{\cdot\cdot}$	$Hg^{\cdot\cdot}$	$Ag^{\cdot}$
1.0×10^{-3}	5.0×10^{-4}	1.2×10^{-4}	1.2×10^{-4}	1.2×10^{-4}	6.0×10^{-5}	1.5×10^{-5}	2×10^{-6}	2×10^{-6}

3.3 Anti-microbial Activity of Silver Nitrate

Silver nitrate that forms silver ions in water solution has strong anti-microbial activity. The activity at much lower concentration than that, causes protein denaturation, has long been known. Consequently, silver nitrate has a history of its use as a therapeutic for bacterial infection, and antiseptic for midwife, nurse and military hospitals. Silver is considered to have inhibitory activities of metabolic functions of respiratory and electron transport systems of microbes, and mass transfer across cell membranes.

The anti-microbial activity of silver ions has been studied in relation to bacterial leaching in mining where silver ion inhibits growth of bacteria useful in leaching[9]. Growth of this sulfur bacteria, *Thiobacillus ferrooxidans*, can be slightly inhibited by silver nitrate at a concentration of 0.1 ppm, and the growth was completely suppressed at 1.0 ppm[10]. As for the mechanism of action, the silver ion is first adsorbed to the surface of microbial cells, incorporated within the cells by means of active transport, inhibiting a range of metabolic enzymes, to demonstrate anti-microbial activity. Since silver ions very easily react with proteins, they may react with different enzyme proteins, after incorporation in the microbial cells, and thus inhibit metabolic processes necessary for sustaining life.

Several systems which produce active oxygen and show anti-microbial activity have been reported. However, the results from experiments with yeast revealed that Ag-Zeolite's anti-microbial activity is observed in both aerobic and anaerobic conditions, and that the degree of activities on yeast were almost the same regardless of the oxygen concentration and the existence of light (Table 5). Thus, the active oxygen does not seem to be directly related in this case[11].

Although silver demonstrates a fairly broad spectrum of anti-microbial activity, some bacteria resistant to silver and those which take up silver into the cell were discovered.

3.4 Anti-microbial Activity of Ag-Zeolite

As mentioned earlier, Zeolite contains, in its crystalline structure, sodium ions that can be substituted by other metallic ions, i.e. the Ag ion can efficiently substitute the Na ion to form Ag-Zeolite. In the manufacture of Ag-Zeolite, synthetic zeolite is normally used.

Table 5. *Growth Inhibitory Effects of High Exchange Ag-Zeolite Concentrations on Saccharomyces cerevisiae under Aerobic and Anaerobic Conditions*

	Incubation	35	30	25ppm	20ppm	15ppm	10ppm	5ppm	0ppm
Ag-40	Aerobic	0	0	1.5×10^0	3.8×10^0	5.3×10^2	7.1×10^2	2.4×10^4	2.4×10^2
	Anaerobic	0	0	1.5×10^0	1.5×10^0	4.1×10^1	6.4×10^2	1.4×10^3	1.4×10^3
Ag-70	Aerobic	0	0	0	0	8.2×10^1	1.4×10^2	6.0×10^3	1.9×10^2
	Anaerobic	0	0	0	0	1.0×10^0	3.7×10^2	1.1×10^3	1.4×10^3
Aj-40	Aerobic	0	0	0	0	3.4×10^2	6.7×10^2	3.0×10^3	1.9×10^2
	Anaerobic	0	0	0	0	1.0×10^0	3.6×10^1	7.4×10^2	1.4×10^3
Aj-70	Aerobic	0	0	0	0	0	0	6.0×10^0	1.9×10^2
	Anaerobic	0	0	0	0	0	0	8.6×10^2	1.4×10^3

Ag :Silver-exchanged, Aj :Silver and Zinc-exchanged.   YM medium was used for incubation. Anaerobic condition was prepared by using excess oxygen absorber

The lower the concentration of nutrient in media, the lower the concentration of Ag ion that is required to demonstrate anti-microbial activity. In a diluted medium to the extent of 500-1000 times, the activity was noticeable even at the concentration of 0.02-0.05 ppm of Ag ion (Table 6). The release of Ag ions from Ag-Zeolite powder is not observed in pure water, but in nutrient media, almost all the Ag are released. It is suggested that the released Ag ions react with sulfur compounds or other active constituents in media, and only a part of this released Ag ions demonstrate anti-microbial activity. When the anti-microbial activity of Ag-Zeolite powder, containing 2.5% of Ag, is determined in nutrient media, Ag at the concentration of 1.5-3.5 ppm suppresses microbial growth. Higher degree of substitution of Ag ion in zeolite increases the activity against microbes.

Table 6. *Effects of Concentrations (ppm) of Ag-Zeolite and Dilution Times of Culture Medium on the Growth of Saccharomyces cerevisiae*

Dilut.	Ag-Z.	Counts	Dilut.	Ag-Z.	Counts	Dilut.	Ag-Z.	Counts	Dilut.	Ag-Z.	Counts
1000	0	2.8×10^3	100	0	1.0×10^6	10	0	2.1×10^7	1	1	1.7×10^8
	1	2.9×10^3		5	1.2×10^4		10	1.4×10^7		100	9.8×10^7
	5	4.0×10^2		10	3.0×10^3		100	no growth		1000	no growth
	10	no growth		100	no growth		1000	no growth		5000	no growth

Zeolite is considered to retain Ag ions in stable and effective conditions, and to facilitate the anti-microbial activity against the microbes that may come into close contact with it. The unique feature of Ag-Zeolite's anti-microbial activity is the highly broad spectrum of bacteria, with little specificity of bacterial genera and it is almost equally effective to bacteria, yeast and mycelium fungi. This fact too, suggests that the action of Ag ions is the inhibitory action to the components or functions that many different microbes share commonly. To the spores of heat-resistant bacteria, it does not demonstrate any activity, while the activity can be only demonstrated to the spores that already germinated.

Ag-Zeolite can maintain its anti-microbial activity only while it retains Ag ions in the skeleton of Zeolite. But, this activity is lost, if all Ag ions get eluted, or if Ag ions get inactivated by reacting with medium constituents to form inert substances. In various foods, there are many substances that react and weaken the activity of Ag ions such as sulfates, hydrogen sulfide, and sulfur-containing amino acids, and therefore, they may weaken the Ag-Zeolite's activity at ambient temperatures. Consequently, the mechanism of the action has to be elucidated, and efficient uses of the effects of inhibitants, as well as the limit of activity of Ag-Zeolite,have to be clarified.

3.5 The Application of Ag-Zeolite to Packaging Resources

Different kinds of anti-microbial packaging materials, in which Ag-Zeolite is incorporated in plastics have been developed and tested for applications. As Ag-Zeolite is expensive, it is laminated as a thin co-extruded layer of 3-6 μm containing Ag-Zeolite (Figure 4), or this film is applied on the surface of formed containers. The anti-microbial activity is demonstrated by the Ag ions contained in zeolite particles on the film surface, and therefore, thicker film may not affect the activity of Ag ions inside the film at all. The amount of Ag-Zeolite added may influence the heat-sealing strength and other physical properties like transparency of the packaging films. The normal incorporation level is 1-3 %, and up to 5 % has been tested.

A variety of food poisoning and putrefying bacteria suspended in physiological saline water were sprayed on the surface of polyethylene and other plastic films containing 1% of Ag-Zeolite, to determine the reduction of viable counts. All bacteria were eliminated in 1-2 days on the Ag-Zeolite containing films (Table 7), where as the control film retained the initial viable counts in most cases. It has become clear that bacteria may disappear quite rapidly when bacteria are closely attached to the films or the films do not have any nutrients.

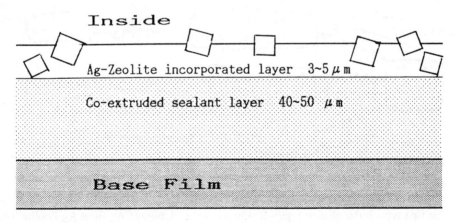

Figure 4. *Structure of laminated film with Ag-Zeolite*

3.6 The Interactions between Anti-microbial Activity and Foods

When a plastic film containing Ag-Zeolite is placed in a shaking flask with culture media, there should be a sort of balance between cell growth rate and death rate of cells in contact with Ag ion, and consequently, the film containing 1-3 % of Ag-Zeolite does not show any anti-microbial activity in nutrient-rich culture media. In order to enhance the anti-microbial activity, the use of plastic materials with high substitution of Ag-Zeolite, increasing its content, or the application of Ag-Zeolite on the film surface to increase the contact surface area, may be another approach that would help.

Table 7. *Anti-microbial Effects of Ag-Zeolite-incorporated Polyethylene Film*

Microorganisms	Sample	0 hr	24 hrs	48 hrs
Escherichia coli	a	1.7×10^5	< 10	< 10
	b	1.5×10^5	5.0×10^6	4.0×10^5
Staphylococcus aureus	a	1.0×10^5	2.6×10^3	< 10
	b	1.1×10^5	4.6×10^4	8.7×10^4
Salmonella typhimurium	a	2.8×10^4	3.2×10^2	< 10
	b	3.6×10^4	3.6×10^6	4.4×10^6
Vibrio parahaemolyticus	a	2.8×10^4	< 10	< 10
	b	1.7×10^4	1.6×10^3	5.6×10^4

a : Polyethylene film incorporated with 1% Ag-Zn Zeolite
b : Polyethylene film without Ag-Zeolite incorporation

The silver substituted in zeolite molecular matrix gets stimulated to become eluted by the amino acids in foods, and this effect was found to be different depending on amino acid involved. There are largely three types of amino acid groups and related compounds on the basis of Ag elution pattern and the influence to anti-microbial effects (Figure 5).

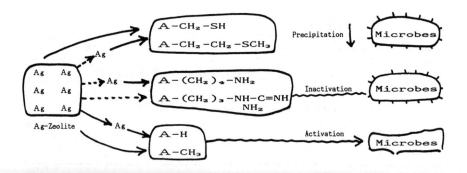

Figure 5. *Effects of amino acid on the anti-microbial activity of Ag-Zeolite*

The first type is the glycine-type. It stimulates the elution of Ag from Ag-Zeolite, but does not interfere in the action of Ag ions as its reaction with Ag is weak. The second type is the lysine-type. The elution of Ag ions are weaker than with glycine, but their association with Ag ions are relatively strong and inhibit the anti-microbial activity. The third is the cysteine-type. Both the elution and association with Ag are strong and extremely inhibitory on anti-microbial activity. The behavior of albumin protein is that of the weak lysine-type, and glutathione of tripeptide with SH-group was similar to the cysteine-type.

From these data, certain kinds of amino acids and proteins are considered to influence on the anti-microbial activity of Ag-Zeolite. Therefore, it would be necessary to evaluate the quality and quantity of amino acids and proteins in foods in terms of anti-microbial activity when films containing Ag-Zeolite is applied to food quality preservation. This would provide important information to determine the substitution ratio and the amount of Ag-Zeolite to be incorporated in the films.

At the same concentration of Ag-Zeolite in films, it may turn out that the effects are positive or negative depending on the kinds of foods (kind and amount of nutrients, salts, pH etc.) or preservation temperatures. When foods are packaged with films containing Ag-Zeolite, it is necessary to evaluate nutrient level of food side, and to set the appropriate Ag contents that meet the nutrient level of foods. In the actual food packaging, the microbial growth may be suppressed on the contact surface with Ag-Zeolite-incorporated film, but can not be suppressed in the area without contact. Therefore, the effects cannot be expected when nutrient-rich foods are packaged at the state of small relative contact areas, whereas very high effects can be expected when nutrient-poor drinks like mineral water or oolong tea are going through low-temperature distribution channel. These good results have been obtained (Table 8,9).

In the course of technical development for applications, the specific properties of Ag-Zeolite that has been described must be well understood.

Table 8. *Anti-microbial Effects of Ag-Zeolite-incorporated Polyethylene Film on the Viable Count of E. coli and Streptococcus pneumoniae in Chinese Oolong Tea*[12]

	Sample		0 hr	5 hrs	24 hrs	48 hrs
2 5 ℃	Ag-Zeolite	Total count	1.7×10^6	4.4×10^5	2.5×10^3	3.3×10^1
		E. coli	3.6×10^5	1.2×10^4	<10	<10
		S. pneumonia	5.4×1^5	1.8×10^3	9.3×10^2	2.1×10^1
2 5 ℃	Control	Total count	1.7×10^6	6.5×10^5	7.6×10^4	3.9×10^7
		E. coli	3.6×10^5	1.9×10^4	5.0×10^3	1.0×10^6
		S. pneumonia	5.4×10^5	2.2×10^3	7.6×10^2	9.4×10^1
1 0 ℃	Ag-Zeolite	Total count	1.7×10^6	3.6×10^5	8.4×10^4	4.0×10^4
		E. coli	3.6×10^5	5.6×10^3	5.2×10^1	<10
		S. pneumonia	5.4×10^5	5.4×10^3	1.2×10^4	2.8×10^3
1 0 ℃	Control	Total count	1.7×10^6	7.1×10^5	1.5×10^5	2.3×10^4
		E. coli	3.6×10^5	5.8×10^3	8.0×10^2	1.6×10^2
		S. pneumonia	5.4×10^5	6.5×10^3	3.2×10^4	2.2×10^4

Total count : PCA medium

Table 9. *Anti-microbial Effects of Ag-Zeolite-incorporated Polyethylene Film on the Viable Count of Pseudomonas aeruginosa in Chinese Oolong Tea*[12]

	Sample	0 hr	5 hrs	24 hrs	48 hrs
2 5 ℃	Ag-Zeolite	9.2×10^5	5.1×10^5	<10	<10
	Control	9.2×10^5	5.5×10^5	2.3×10^4	<10
1 0 ℃	Ag-Zeolite	9.2×10^5	7.9×10^5	2.5×10^5	2.6×10^2
	Control	9.2×10^5	8.1×10^5	2.6×10^5	8.6×10^4

Medium : Plate count agar

References

1. T. Ishitani, et. al.: Handbook for Functional Packaging for Food, 1989, 14-22
2. K.Kazuki, K.Mita, K.Yoshida and T.Ishitani: *Packaging Research*, 1990, **10**, 11
3. K.Kazuki, K.Mita, K.Yoshida and T.Ishitani: *Packaging Research*, 1990, **10**, 21
4. Y.Osajima, T. Matsui: *J. Packaging Science and Technology*, 1994, **3** (1) 3
5. T. Ishitani : *Packaging Japan*, 1993, **14**,(9) 30
6. H.Nishino and J.Yoshii: *Food Science, Japan*, 1988, **30**,1
7. K.Tokuoka, Y.Mihara, J.Kamida and T.Ishitani :*Nippon Shokuhin Kogyou Gakkaishi*, 1981, **38**, 1111
8. T. Ishitani : Advance in Food Technology (IV), Korin, Tokyo, 1990
9. R.C.Tilton et al.: *Appl. Environ. Microb.*, 1978, **35**, 1116
10. T. Sugio: *Agric. Biol. Chem.*, 1981, **45**, 2037
11. Y. Ishikawa, T. Ishitani: *Nippon Shokuhin Kogyo Gakkaishi*, 1988
12. Japanese Center for Food Analysis : Antimicrobial Packaging Materials, 1992

Active Packaging for Fresh Produce

Brian P. F. Day

DEPARTMENT OF PRODUCT AND PACKAGING TECHNOLOGY, CAMPDEN FOOD AND DRINK RESEARCH ASSOCIATION, CHIPPING CAMPDEN, GLOUCESTERSHIRE GL55 6LD, UK

1. INTRODUCTION

The depletion of oxygen (O_2) and enrichment of CO_2 are natural consequences of the progress of respiration when fresh fruit or vegetables are stored in an hermetically sealed package or container. Such modification of the atmospheric composition results in a decrease in the respiration rate of plant material.[1,2]

If produce is sealed in an impermeable film, in-pack O_2 levels will fall to very low concentrations where anaerobic respiration will be initiated. Figure 1(a) schematically illustrates this scenario. Anaerobiosis, with its accumulation of ethanol, acetaldehyde and organic acids, is usually associated with undesirable odours and flavours and a marked deterioration in product quality. In addition, there is a risk of the growth of anaerobic pathogens, such as *Clostridium botulinum*. Therefore, a minimum level of 2-3% O_2 is recommended to ensure that potentially hazardous conditions are not created.[3]

Conversely, if fruit or vegetables are sealed in a film of excessive permeability, little or no atmospheric modification will result within the package. Figure 1(b) schematically illustrates this scenario. In addition, moisture loss will cause undesirable wilting and shrivelling, and therefore fully permeable films are unsuitable for fresh produce packaging.

However, if film of correct intermediary permeability is chosen, a desirable equilibrium modified atmosphere (EMA) is established when the rates of O_2 and CO_2 transmission through the package equals the product's respiration rate. Figure 1(c) schematically illustrates this scenario. The exact EMA attained will obviously depend on the product's intrinsic respiration rate but will also be greatly influenced by various extrinsic factors (section 3). These factors need to be optimised for each commodity so that the full benefits of MAP can be realised.

1.1 Controlled Atmosphere Storage Vs. MAP

Similar to MAP, controlled atmosphere storage (CAS) refers to the storage of food in an atmosphere that is different from the normal composition of air. In CAS the atmospheric components are precisely adjusted to specific concentrations; however, in MAP there is no way of controlling atmospheric components at specific concentrations once a package has been hermetically sealed.[4]

CAS is used for the warehouse storage of whole fruit and vegetables or the bulk CA road or sea-freight transport of perishable foods. For example, CAS is used for the long-term storage of apples, pears, kiwi fruit and cabbage.

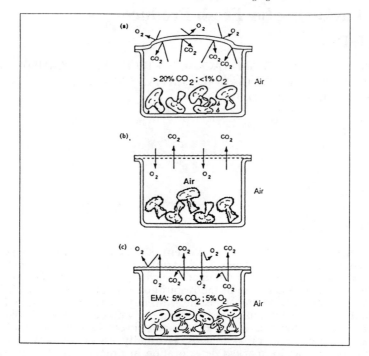

Figure 1 - Schematic Representation of the Three Packaging
Scenarios of MA Packed Produce

(a)   Barrier film: undesirable anaerobic conditions
(b)   Fully permeable film: no desirable atmospheric modification
(c)   Intermediate permeable film: desirable EMA

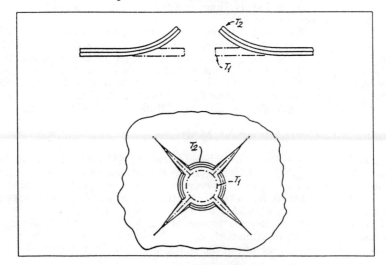

Figure 2 - Temperature Sensitive Apertures Formed From
A Bilayer Laminated Material with Star-Shaped Holes or Petals
$(T_1 = 5°C, T_2 = 20°C)$

Such technology has been established for over 60 years, and since then more than 4000 research articles have been published which identify the optimal CAS conditions for each commodity and cultivar.[5,6] It is not within the scope of this paper to review CAS technology.

1.2 Advantages and Disadvantages of MAP

The advantages and disadvantages of MAP for fresh produce have been extensively reviewed.[1,6] Depleted O_2 and/or enriched CO_2 levels can reduce respiration, delay ripening, decrease ethylene production and sensitivity, retard textural softening, slow down compositional changes associated with ripening, reduce chlorophyll degradation and enzymic browning, alleviate physiological disorders and chilling injury, maintain colour, and preserve vitamins of fresh produce, thereby resulting in an extended quality shelf-life.[1,5] The effects of depleted O_2 and enriched CO_2 levels on respiration and fruit ripening are additive and can be synergistic [5]. However, exposure of fresh produce to O_2 or CO_2 levels outside the limits of tolerance (section 1.4) for a particular commodity can initiate anaerobic respiration with the production of undesirable odours and flavours, as well as cause other physiological disorders.

1.3 Methods of Creating MA Conditions

MAs can be created either passively by the commodity or intentionally via active packaging.[5]

1.3.1 Passive MA. MAs can passively evolve within an hermetically sealed package as a consequence of a commodity's respiration, i.e. O_2 consumption and CO_2 evolution. If a commodity's respiration characteristics are properly matched to film permeability values, then a beneficial MA can be passively created within a package. If film of correct intermediary permeability is chosen, then a desirable EMA is established when the rates of O_2 and CO_2 transmission through the package equal a product's respiration rate (section 1). It is important not to select films of insufficient permeability because of the hazard of creating anaerobic conditions and/or injuriously high levels of CO_2.

1.3.2 Active packaging. In contrast to CAS, in MAP there is a limited ability to regulate a passively established MA within an hermetically sealed produce package. There may be circumstances when it is desirable to actively establish and adjust the atmosphere within a produce package, and this can be done by utilising active packaging techniques.

By pulling a slight vacuum and replacing the package atmosphere with a desired mixture of CO_2, O_2 and N_2, a beneficial EMA may be established more quickly than a passively generated EMA. For example, recent research has shown that the sensory appearance of mixed lettuce (30% Iceberg, 30% Chinese leaf, 20% Lollo Rosso and 20% endive) was improved by gas flushing with 5% O_2, 5% CO_2 and 90% N_2 compared to air packing. Enzymic browning was visible at a later storage time in the gas flushed samples since the rate of browning is partially dependent on O_2 concentration.[2]

Another active packaging technique is the use of O_2, CO_2 or ethylene scavengers/emitters.[6,7] Such scavengers/emitters are capable of establishing a rapid EMA within hermetically sealed produce packages. However, the use of O_2 scavengers within high moisture produce packages would exacerbate the development of undesirable anaerobic conditions and hence is not recommended.[4]

In addition, ethylene scavengers can help to ensure the delay of the characteristic rise in respiration rate of climacteric fruit [8]. Also, CO_2 scavengers can prevent the build-up of CO_2 to injurious levels, an undesirable situation that can occur for some commodities during passive modification of a produce package atmosphere.[5,7,9]

1.4 Optimal Equilibrium

The limits of tolerance to low O_2 and high CO_2 levels, outside which physiological damage occurs, are subject to several variables, such as type of produce, cultivar, temperature, physiological condition, maturity and previous treatment.[5]   Since MA conditions generated within a package may fluctuate slightly, the practical "optimal" atmosphere should be one that is not too close to an injurious EMA. As a generalisation, EMAs containing between 2-5% O_2 and 3-8% CO_2 have been shown to extend the shelf-life of a wide variety of fruit and vegetables.[1] However, it is understood that a handful of packaging companies are experimenting with high O_2/high CO_2 (e.g. 70% O_2/30% CO_2) for certain prepared green vegetables with surprisingly good results. Also, argon has been shown to inhibit the tissue fermentation of sliced tomatoes and extend the shelf-life of prepared fruit.[10]   These novel potential gas mixture applications need further investigation.

2. BACKGROUND INFORMATION

2.1 Factors Affecting Shelf-life

Shelf-life may be defined as the period of time from harvest or manufacture to consumption that a food product remains safe and wholesome under recommended production and storage conditions.[4] With respect to MA packed fruit and vegetables, shelf-life is affected by numerous intrinsic properties of fresh produce as well as various extrinsic factors.[11]

2.2 Intrinsic Properties of Fresh Produce

2.2.1 Respiration rate. Generally which compositional changes are taking place within plant material, and hence indicates the potential shelf-life of individual fruit or vegetables.[1] Table 1 classifies fresh fruit and vegetables according to their respiration rate and degree of perishability. Those fruit or vegetables with an extremely high respiration rate and correspondingly extremely high degree of perishability will have a very short shelf-life, e.g. sprouting broccoli, peas in pod, sliced mushrooms and Julienne-cut carrots.

2.2.2 Acidity (pH). The pH of individual fruits or vegetables will influence the types of spoilage and food poisoning microorganisms liable to grow during a commodity's shelf-life. Most fruits such as lemons, oranges, pineapples, apples and peaches have pH values below 4.5.[12] Under such acidic conditions, *Clostridium botulinum* cannot grow and produce its potentially deadly toxin. Consequently, such acidic fruits can be safely vacuum packed or packaged in hermetically sealed films of low permeability so that anaerobic conditions can be passively generated.   Conversely, most vegetables such as lettuce, carrots, potatoes, mushrooms, broccoli and beansprouts have pH values above 4.5[12], and consequently *Clostridium botulinum* is able to grow when such commodities are stored under anaerobic conditions. The only barrier to *Clostridium botulinum* growth is then storage at temperatures below 3°C.[4]

Table 1 - Classification of Selected [a] Fruit and Vegetables According to Their Respiration Rate and Degree of Perishability in Air and 3% O^2

Commodity	Respiration rate - CO_2 production $(ml\ kg^{-1} h^{-1})$[b]						Relative respiration rate at 10°C in air
	In air			In 3% O_2			
	0°C	10°C	20°C	0°C	10°C	20°C	
Onion (Bedfordshire Champion)	2	4	5	1	2	2	
Cabbage (Decema)	2	4	11	1	3	6	
Beetroot (storing)	2	6	11	3	4	6	Low
Celery (white)	4	6	19	3	5	12	<10
Cucumber	3	7	8	3	4	6	
Tomato (Eurocross BB)	3	8	17	2	3	7	
Lettuce (Kordaat)	5	9	21	4	6	14	
Peppers (green)	4	11	20	5	7	9	
Carrots (whole, peeled)	—	12	26	—	—	—	
Parsnip (Hollow Crown)	4	14	23	3	6	17	Medium
Potatoes (whole, peeled)	—	14	33	—	—	—	10-20
Mango	—	15	61	—	—	—	
Cabbage (Primo)	6	16	23	4	8	17	
Lettuce (Kloek)	8	17	42	8	13	25	
Cauliflower (April Glory)	10	24	71	7	24	34	
Brussels sprouts	9	27	51	7	19	40	High
Strawberries (Cambridge Favourite)	8	28	72	6	24	49	20-40
Blackberries (Bedford Giant)	11	33	88	8	27	71	
Asparagus	14	34	72	13	24	42	
Spinach (Prickly True)	25	43	85	26	46	77	
Watercress	9	43	117	5	38	95	Very high
Broad beans	18	46	82	20	29	45	40-60
Sweetcorn	16	48	119	14	32	68	
Raspberries (Malling Jewel)	12	49	113	11	30	73	
Carrots (Julienne-cut)	—	65	145	—	—	—	
Mushrooms (sliced)	—	67	191	—	—	—	Extremely high
Peas in pod (Kelvedon Wonder)	20	69	144	15	45	90	>60
Broccoli (sprouting)	39	91	240	33	61	121	

[a] Unless stated, produce is whole and unprepared.
[b] mg CO_2 converted to ml CO_2 using densities of CO_2 at 0°C = 1.98, 10°C = 1.87, 20°C = 1.77.

2.2.3 Water Activity (a$_w$). Fresh fruit and vegetables are high moisture foods which have a$_w$ values between 0.95 and 1.00. Under such high a$_w$ conditions, spoilage and food poisoning microorganisms can readily grow. Consequently, food technologists need to rely on the combination of optimal chilled temperatures and MAP to inhibit the growth of such microorganisms so as to extend the shelf-life and ensure the safety of packaged fresh produce.

2.2.4 Biological structure. The resistance of plant tissues to diffusion of O_2, CO_2, ethylene and water vapour is dependent on the biological structure of indfvidual fruit or vegetables. Resistance to gas diffusion influences a commodity's tolerance to depleted levels of O_2 and elevated levels of CO_2 (section 1.4). Resistance to gas diffusion varies depending on the type of commodity, cultivar, part of the plant, severity of preparation, and stage of maturity, but appears to be little affected by temperature.

3. EXTRINSIC FACTORS TO OPTIMISE

3.1 Harvesting

Harvesting fruit and vegetables at optimal maturity is the most basic factor affecting the quality and subsequent shelf-life of produce.[13] Determination of optimal maturity depends on many factors, including the intended use of the produce. Generally speaking, fruit and vegetables intended for processing and packaging are normally harvested prior to their peak maturity. At this stage their texture is firmer, and thereby mechanical damage during handling and processing can be minimised.

3.2 Handling

Minimising mechanical injury is one of the main factors affecting the quality and shelf-life of produce.[13] Careful handling resulting in reduced bruising can increase the shelf-life of produce dramatically and also help to minimise produce waste.

3.3 Hygiene

Rigorous and systematic control of hygienic practices is essential during harvesting, handling, preparation, processing, packaging, storage, distribution, and final consumption of MA packed fruit and vegetables.[4] Strict conditions of hygiene must be maintained to prevent cross-contamination with food poisoning bacteria.

3.4 Temperature

Maintaining proper temperature control after harvesting is one of the most important extrinsic factors affecting the quality of MA packed produce. The best practice is to harvest early in the morning or at night and remove field heat as soon as possible by various chilling methods, such as air blast chilling, hydrocoolers or vacuum coolers.[12] Temperatures in the range 0-5°C are generally chosen for storage and distribution of most MA packed produce. At such chilled temperatures, respiration rates are significantly lowered and the growth of spoilage and food poisoning microorganisms is restricted.

3.5 Water Loss and Relative Humidity (RH)

Loss of moisture, with consequent wilting and shrivelling, is one of the obvious ways in which freshness of fruit and vegetables is lost.[13] Since fruit and vegetables are 80-95% water, they lose moisture rapidly whenever the RH is less than 80-95% below saturation. Moisture losses of 3-6% are usually enough to cause marked deterioration of quality for many kinds of produce. Consequently, it is important to reduce such moisture losses by lowering temperature, raising RH and reducing air movement. All of these methods for reducing moisture losses of fresh produce can be accomplished by MAP.

Most hydrophobic plastic films used for MAP of fresh produce are relatively good water vapour barriers and are able to maintain high in-pack humidities, even when conditions are dry in the external atmosphere. However, there is the problem that the in-pack RH can get too high, thereby causing moisture condensation and conditions favourable for microbial growth, resulting in decay of the commodity.[5]

3.6 Packaging Materials

The main characteristics to consider when selecting packaging materials for MAP of fruit and vegetables are gas permeability, water vapour transmission rate, mechanical properties, type of package, transparency, sealing reliability, and microwaveability.

As mentioned in section 1, packaging films of correct intermediary permeability are usually chosen for the MAP of respiring fruit and vegetables. Table 2 lists the oxygen and water vapour transmission rates of a wide variety of packaging films that may be used in the MAP of fresh produce. Using these types of film, desirable EMAs can be developed. However, due to differences in the respiration rates of individual fruit or vegetables and the effect of temperature on both respiration and gas permeability, the type of packaging film required to achieve any specific EMA must be defined for each commodity at one specific storage temperature.[1]

For highly respiring produce such as mushrooms, beansprouts, leeks, herbs, peas and broccoli, traditional films like LDPE, PVC, EVA and OPP are generally not permeable enough. Newly developed highly permeable microperforated films such as "P-Plus" (Sidlaw Packaging, Bristol, UK) and "Crop" (Allen Packaging, Norfolk, UK) appear to be most suitable at the present time, but future new developments may change this picture.[10]

The gas permeability of a particular packaging material depends on several factors such as the nature of the gas, the structure and thickness of the material, temperature, and RH. CO_2, O_2 and N_2 permeate at quite different rates. However, the order $CO_2 > O_2 > N_2$ is always maintained and the ratios pCO_2/pO_2 and pO_2/pN_2 are usually in the range 3-5. Hence, it is possible to estimate the permeability of a plastic material to CO_2 or N_2 when only the O_2 permeability is known.[4]

Although the information in Table 2 is a useful guide to appropriate film selection, further information on O_2 and CO_2 permeabilities of plastic films at more realistic chilled temperatures and high RHs is required for accurate selection of film to match a particular commodity's respiration rate.[1] With this information it would be possible to predict the required film to attain a specific optimised EMA for a given commodity.

Table 2 - Oxygen and Water Vapour Transmission Rates of Selected[a] Packaging Materials for Fruit and Vegetables

Packaging film (25 μm)	Oxygen transmission rate ($cm^3\ m^{-2}\ day^{-1}\ atm.^{-1}$) 23°C:0% RH[b]	Relative permeability at 23°C:0% RH	Water vapour transmission rate ($g\ m^{-2}\ day^{-1}$) 38°C:90% RH[b]	Relative water vapour transmission rate at 38°C:90% RH
Aluminium (Al)	<0.1[1]	Barrier	<0.1[1]	Barrier, <10
Ethylene-vinyl acetate (EVOH)	0.2–1.6[2]	<50	24–120[2]	Variable
Polyvinylidene chloride (PVdC)	0.8–9.2		0.3–3.2	Barrier, <10
Modified nylon (MXDE)	2.4[2]		25	Semi-barrier, 10–30
Polyester (PET)	50–100	Semi-barrier 50–200	20–30	Semi-barrier, 10–30
Polyamide (nylon) (PA6)	80[2]		200	Very high, 200–300
Modified polyester (PETG)	100		60	Medium, 30–100
Metallised orientated polypropylene (MOPP)	100–200[1]		1.5–3.0[1]	Barrier, <10
Unplasticised polyvinyl chloride (UPVC)	120–160		22–35	Variable
Polyvinyl chloride (plasticised) (PVC)	2000–5000[3]	Medium 200–5000	200[3]	Very high, 200–300
Orientated polypropylene (OPP)	2000–2500		7	Barrier, <10
High density polyethylene (HDPE)	2100		6–8	Barrier, <10
Polystyrene (PS)	2500–5000		110–160	High, 100–200
Orientated polystyrene (OPS)	2500–5000		170	High, 100–200
Polypropylene (PP)	3000–3700		10–12	Semi-barrier, 10–30
Polycarbonate (PC)	4300		180	Very high, 100–200
Low density polyethylene (LDPE)	7100	High 5000–10000	16–24	Semi-barrier, 10–30
Polyvinyl chloride (highly plasticised) (PVC)	5000–10000[3]	High 5000–10000	200[3]	Very high, 200–300
Ethylene-vinyl acetate (EVA)	12000	Very high 10000–15000	110–160	Very high, 100–200
Microperforated (MP)	>15000[4]	Extremely high >15000	Variable[4]	Extremely high, >300
Microporous (MPCR)	>15000[4]		Variable[4]	Extremely high, >300

[a] It should be noted that most packaging films for fresh produce are not single films but laminates and co-extrusions.
[b] It should be noted that conditions of O_2 and water vapour transmission rate measurements are not at realistic chill conditions.
[1] Dependent on pinholes.
[2] Dependent on moisture.
[3] Dependent on moisture and level of plasticiser.
[4] Dependent on film and degree of microperforation or microporosity.

Two novel temperature compensating films for produce have been reported recently.[14,15] The Landec Corporation (Menlo Park, California) have engineered "Intelimer" films with a temperature switch point at which the film's permeation changes abruptly and dramatically.   Hence, "Intelimer" films for produce packaging would change in permeation rates at a preset temperature to match or exceed the respiration rate of any fresh produce. The switching mechanism is accomplished by using Landec's patented long chain fatty alcohol based polymeric side chains. Below a preset temperature, these side chains are crystalline and provide a relative gas barrier. However, above this preset temperature, the side chains reversibly convert to an amorphous structure which is up to 1,000 times more permeable to gas and may be tailored to the very large increase in respiration rates of fresh produce at higher temperatures above $5°C$. "Intelimer" films are intended to be used for packaging highly respiring produce but commercialisation is still several years away.[14]

Another temperature compensating film for produce has recently been patented and relies on the same principle as a bimetallic strip but uses two different plastic films instead of two metals (Challis and Bevis, 1992). Temperature sensitive apertures are cut into laminated plastic film in the form of small strips or "petals". At a selected temperature, the laminated film is flat and gives minimal opening areas (Figure 2). However, at higher temperatures the free area for gas exchange increases as the petals curl and the permeability of the film can be engineered to increase in proportion to the rise in respiration rates of produce. The rate of curl is a function of the coefficient of expansion of the two separate films and the extent of opening is controlled by the length and shape of the "petals" surrounding the aperture. Tests on these films are proposed to take place at Pira International and Campden Food and Drink Research Association in the near future to demonstrate the effectiveness of this design concept and to develop commercial opportunities for fresh produce packaging.[16]

Packaging materials used for MAP of fruit and vegetables must have sufficient strength to resist puncture, withstand repeated flexing, and endure the mechanical stresses encountered during handling and distribution.   Poor mechanical properties can lead to pack damage and loss of in-pack atmospheres.

The type of package used will depend on the type of produce to be MA packed and whether the produce is destined for the retail or catering trade. Options include flexible pillow packs, semi-rigid tray and lidding film systems, and bag-in-box containers.[4]

For most MA packed produce, a transparent package is desirable so that the product is clearly visible to the consumer.   However, high moisture produce stored at chilled temperatures has the tendency to create a fog on the inside of the package, thereby obscuring the product. Consequently, many MAP films are treated with coatings or additives to impart antifog properties so as to improve visibility.   These treatments only affect the wettability of the film and have negligible effects on the permeability properties of the film.

MA packs of fresh produce are hermetically sealed, and therefore it is essential that an integral seal is formed in order to maintain an EMA within the pack. Consequently, it is important to select the correct heat sealable packaging materials and to control the sealing operation.

The possibility of using MAP trays directly in microwave ovens for subsequent cooking of fresh vegetables has recently been of commercial interest.   The advantages of such a concept are clearly apparent from a convenience point of view. Fresh whole or prepared vegetables could be MA packed in an appropriate tray and lidding film system to achieve an extended shelf-life.   Then, after perforating the lidding film to allow release of subsequent steam, the tray would be placed directly into a microwave oven for cooking.

Since fresh vegetables have a very high moisture content, cooking times are very rapid. Also, the temperatures reached rarely exceed 100°C, and consequently common inexpensive plastic trays, except PVC/LDPE and PS, can be used since they can withstand such temperatures. Appropriate pillow pack materials could also be used for microwave heating of MA packed vegetables.

3.7 Packaging Machinery

Packaging machinery is another critical parameter that needs careful attention. Generally, the MAP systems used for the retail packaging of fruit and vegetables include vertical form-fill-seal (VFFS) machines to produce flexible pillow packs and thermoform-fill-seal (TFFS) machines to produce semi-rigid tray and lidding film systems. These latter systems utilise thermoformed plastic or composite board, preformed plastic or plastic coated pulp trays in combination with an appropriate lidding film. For catering sized MA packages of produce, bag-in-box systems are very popular.

The type of packaging machinery chosen will influence the seal integrity of MA packs of fresh produce. Poor seal integrity can be due to contamination of the sealing surfaces, poor alignment of the sealing heads, or possible faults in the packaging film itself. Consequently, proper machinery maintenance and quality assurance checks need to be carried out at regular intervals.[4]

3.8 Gas/Product Ratio

Another important parameter for food manufacturers to consider is the gas volume/product volume ratio. To be effective, the gas atmosphere must completely surround the food product in order to extend shelf-life. Generally speaking, in most MAP applications the gas volume/product volume ratio is approximately in the range 3/1 to 1/1. Food manufacturers should investigate the shelf-life implications of product packed under different gas volume/product volume ratios during shelf-life evaluation trials.[4]

Fill weight of produce, pack volume and film surface area will all affect the EMA established within an hermetically sealed package. The respiration rate and fill weight of produce will determine the O_2 demand within the package, while the gas permeability and surface area of film will determine the rates of O_2 and CO_2 transmission into and out of the package. The pack volume is important since it determines the time required before an EMA is established. Ideally, this time should be as short as possible. If gas flushing is used to actively establish a rapid EMA, the volume of gas introduced into a pack must be controlled and should not be too high so as to significantly reduce the packing density during distribution or give the consumer the impression of an undesirable blown pack.

For a given fill weight and pack volume, pillow packs have a larger effective surface area for gas exchange than tray and lidding film packs. For this reason lidding films usually need to be of higher permeability than pillow pack plastic films. Other factors that may affect the EMA established within produce packs include improper stacking, which would restrict the free flow of air surrounding the packs, printing of film and attachment of labels, which may also lower the total gas exchange possible.[1]

4. CONCLUSIONS

The European market for MA packed fresh produce is substantial, with France and the UK leading the way.[10,17] Although MAP has been primarily used for red meats, tremendous opportunities exist for the MAP of fresh produce, especially in the under-developed markets of Germany, Italy, Spain and the Benelux countries. The success of the MA packed vegetable market in France and the UK will stimulate future growth in the MAP of fresh produce in other European nations, the Pacific Rim countries and the USA.

However, the capital cost of MAP equipment and the expense of an adequate chill chain distribution network must also be considered. Increased packaging costs and package sizes may increase distribution costs and limit storage space, but these disadvantages can be offset against the extended shelf-lives achieved, thereby reducing the need for frequent distribution drops. Since MAP will be more expensive than other packaging, it is likely that only higher value crops and value-added prepared produce will be able to absorb the additional costs involved.

Needless to say, there are many advantages of using MAP for fresh produce and these have been highlighted in this paper. The most obvious advantage is the extension of shelf-life without the use of artificial preservatives. Increased shelf-life allows longer distribution lines, reduced wastage, and improved product image, along with the ability to present prepared convenient food items with remaining storage life for the consumer. As with any packaging or processing system, provided the limitations of the system are recognised and the numerous intrinsic properties and extrinsic factors involved are optimised, then the market potential for MA packed fresh produce is enormous.

References

1. Day, B.P.F. (1988) Optimisation of parameters for modified atmosphere packaging of fresh fruit and vegetables. *CAP '88*, Schotland Business Research Inc., Princeton, New Jersey, USA, pp147-170.
2. Day, B.P.F. (1989a) Modified atmosphere packaging of selected prepared fruit and vegetables. *Technical Memorandum No. 524*, Campden Food and Drink Research Association, Chipping Campden, Glos, UK, pp1-65.
3. Bernard, W.J. (1987) Produce packaging to avoid anaerobiosis and prolong quality shelf-life. *CAP '87*, Schotland Business Research Inc., Princeton, New Jersey, USA, pp255-263.
4. Day, B.P.F. (1992a) Guidelines for the manufacture and handling of modified atmosphere packed food products. *Technical Manual No. 34*, Campden Food and Drink Research Association, Chipping Campden, Glos, UK.
5. Zagory, D. and Kader, A.A. (1988) Modified atmosphere packaging of fresh produce. *Fd Technol.* 42 (9), 70-77.
6. Kader, A.A., Zagory, D. and Kerbel, E.L. (1989) Modified atmosphere packaging of fruits and vegetables. *CRC Crit. Rev. Fd Sci. Nut.* 28 (1), 1-30.
7. Day, B.P.F. (1991) Active packaging. In: Proceedings of the *Shelf Life '91* Conference, The Packaging Group Inc., Milltown, New Jersey, USA.
8. Szikla, Z. and Zsoldas, B. (1993). Ethylene absorbing paper for the packaging of fresh fruits and vegetables. *MAPack '93*, Institute of Packaging Professionals, Herndon, Virginia, USA.
9. Labuza, T.P. and Breene, W.M. (1989) Applications of active packaging for improvement of shelf-life and nutritional quality of fresh and extended shelf-life foods. *J. Fd Processing Preserv.* 13, 1-69.

10. Day, B.P.F. (1992b) An update on CA/MA/vacuum packaging developments in Europe. *Foodplas '92* Conference, The Plastics Institute of America, Fairfield, New Jersey, USA.

11. Day, B.P.F. (1989b) Extension of shelf-life of chilled foods. *European Food and Drink Review*, Autumn, 47-56.

12. Holdsworth, S.D. (1983) The preservation of fruit and vegetable food products. *Science in Horticulture Series*, L. Broadbent (Ed.), MacMillan Press, London, UK, pp61-98.

13. Zomorodi, B. (1990) The technology of processed/prepackaged produce. Preparing the product for modified atmosphere packaging (MAP). *CAP '90*, Schotland Business Research Inc., Princeton, New Jersey, USA, pp301-317.

14. Anon. (1992). Temperature compensating films for produce. *Prepared Foods*. Sept. edition, 95.

15. Challis, A.A.L. and Bevis, M.J. (1992). Material having a passage therethrough. *International Patent Application Number WO92/00537*.

16. Barnetson, A. (1993). Research proposal - A temperature responsive packaging system for fruit and vegetables. *Pira International*, Leatherhead, Surrey, UK.

17. Day, B.P.F. (1990) A perspective of modified atmosphere packaging of fresh produce in western Europe. *Fd Sci. Tech. Today* 4 (4), 215-221.

Trends and Applications of Active Packaging Systems

J. Miltz, N. Passy, and C.H. Mannheim

DEPARTMENT OF FOOD ENGINEERING AND BIOTECHNOLOGY,
TECHNION-ISRAEL INSTITUTE OF TECHNOLOGY, HAIFA, 32000, ISRAEL

1 INTRODUCTION

The main objectives of food packages are to protect the contents and provide maximum shelf life. Until recent years, the general attitude was that no or minimum package-product interaction should occur. However, new technologies are emerging in which the food, package and the environment inside the package interact in a positive way resulting in either an extension of shelf-life or improved product characteristics. These technologies, called "Active Packaging" (AP), combine advances in Food Technology, Biotechnology, Packaging, Material Science, and comply with consumer demands for high quality and "fresh like" products [1].

The environmental factors affecting shelf life of foods include oxygen, carbon dioxide, moisture and in fresh produce also ethylene. The technologies involved in AP systems include: oxygen scavenging, carbon dioxide and ethylene control and the use of moisture regulators, chemical agents, application of ethanol and other means [1].

1.1 ACTIVE PACKAGING SYSTEMS

1.1.1 Oxygen scavenging. Oxygen causes oxidation of lipids in foods, loss of nutrients, discoloration, and browning. Oxygen scavenging from the package atmosphere by chemical (e.g. iron compounds) or enzymatic systems (e.g. glucose oxidase-catalase) serves to prevent these undesirable reactions. Oxygen's scavengers include the following materials:

A. Iron compounds. These absorbers are metallic reducing agents that comprise (most commonly) of powdered iron oxide and ferrous compounds which are oxidized to the ferric state when in contact with oxygen. They are combined with various catalysts to initiate the reaction. Non metallic and organo-metallic compounds have also been developed [2]. The absorbers are packed in small sachets (packets), and are placed inside the food package. The rate of oxidation of the food, the residual oxygen in the pack and the permeability of the package to oxygen, determine the amount of iron needed for a specific package[1]. According to the manufacturers, removal of oxygen is complete within 0.5-4 days (depending on above parameters)[3]. The manfacturers also state that the sachets can keep oxygen levels of less than 0.1% for a few months. Oxygen absorbers were found to be more effective than nitrogen flushing, or vacuum packaging, in preventing oxidation of fat containing products such as peanuts.

     B. Other Oxygen Controlling Systems. Another system which acts as an oxygen scavenger, is the CMB OXBAR system[4]. It is composed of a PET/MXD6/Co film, where the PET serves as the structural material and the active ingredients are MXD6 nylon and a cobalt salt. The cobalt catalyses the reaction of MXD6 with the diffusing oxygen. The company claims that the system is independent of temperature and humidity and has a long life. It can be used in beer and juice bottles.

     Another way of controlling oxygen level in a food package is to use the Glucose Oxidase enzyme which reacts with the substrate and scavenges oxygen. One idea to apply this enzyme is to bind it to the inside surface of a film enabling it to react with excess oxygen[1].

     1.1.2 Carbon dioxide control. CO_2 extends shelf life of foods due to its microstatic activity, and of some fresh produce by reducing respiration. The gas concentration can be controlled by passive or active means. Addition of the gas can be achieved by introducing it into the package prior to sealing or by generating it during storage; its reduction can be accomplished by absorption or by permeation through the package.

     Similarly to oxygen, a sachet containing a material that can scavenge or generate carbon dioxide can be introduced into the package. Some systems exist which absorb oxygen and emit carbon dioxide at the same time[2]. Systems which remove CO_2 contain calcium hydroxide which reacts with the carbon dioxide (at sufficiently high humidities) producing calcium carbonate. Its application is primarily in preventing bursting of coffee packages by the high levels of CO_2 generated after roasting.

     1.1.3 Ethylene control. Ethylene produced during respiration of fruits and vegetables, especially climacteric ones, causes some undesirable effects which include[5]: accelerated senescence and loss of green color in some immature fruits and leafy vegetables; accelerated ripening of fruits during handling and storage; phenomena such as: russet spotting on lettuce; formation of a bitter compound in carrots; sprouting of potatoes; toughening of asparagus; reduced quality of flowers; physiological disorders in flowering bulbs, etc.

     Ethylene can be removed by ventilation, absorbed by a number of chemicals such as potassium permanganate and activated charcoal, or destroyed by ozone.

1.1.4 Moisture Regulators. Absorption of moisture by hygroscopic food products may cause a deterioration in their quality. In other cases, loss of moisture is undesirable due to quality or economic reasons. Barrier films are used to protect packed foods against moisture penetration (or loss) through the package. In some cases a desiccant is placed in the package to absorb excess moisture. The most common desiccant is the non-toxic and non-corrosive silica gel.

1.1.5 Ethanol vapor release. Ethanol was shown to inhibit microbial growth when sprayed onto the surface of foods, prior to packaging . It was found that the lower the water activity, the less alcohol was needed for this purpose[2].

1.1.6 Chemical agents. Chemical agents, such as antioxidants and UV stabilizers, can be released from or contained in the package and slow down deteriorative reactions in food. Chemical preservatives can be incorporated into or coated on the packaging materials. Carboxy Methyl Cellulose (CMC) containing sorbic acid was coated on a greaseproof paper and used to extend shelf life of bread. Shelf life extension was achieved by encasing the food with edible films that contained sorbate[7].

1.1.7 Other technologies. Additional technologies that fall under the category of "Active Packaging" include[2]:

a) Photosensitive dyes impregnated into ethyl cellulose film. When exposed to U.V.light, the dye reacts with oxygen and removes it from the package without the need of a sachet[8] .

b) Anaerobic systems using palladium or platina, in the presence of hydrogen, which converts and binds oxygen to water.

c) Far infra-red radiation in combination with ceramics, silicone and nitride carbide in combination with heat to kill bacteria.

d) Surface treatment of films to change their permeabilities.

e) Incorporation of light absorbers and different minerals into the packaging film to absorb ethylene, CO_2, and odors.

1.1.8 Interactive packages. Another kind of package which can be considered as an "Active Packaging" system (also known as an interactive package) involves the use of components in the package which assist in achieving crisping and browning in the microwave oven. These components are called "susceptors" or thin layer susceptors.

2 EXAMPLES OF FOOD PACKAGE INTERACTIONS AND APPLICATIONS OF ACTIVE PACKAGING SYSTEMS

2.1 Negative Food Package Interactions. Migration of low molecular compounds like monomers, oligomers, additives and residual solvents (from lamination and printing processes) from the package into the contained food is a negative food-package interaction. An additional negative interaction occurs when the product affects the properties of the package.

Styrene migration from polystyrene (PS) packages, was found to impart an off flavor to dairy products[9]. The threshhold taste level of styrene in sour cream was 5 ppb, significantly lower than the migration levels from most PS packages into food simulants[9]. Other studies showed that the amount of residual solvents from printing and lamination processes was much greater in poor pouches and imparted off-flavor to juice, as compared to good ones[10]. The interaction between orange juice and the polyethylene (PE) liner in carton packages was investigated, and it was shown that this liner absorbed a large amount of the orange oil and accelerated ascorbic acid degradation and browning[11]. Absorption of d-limonene by PE film was found to impair its modulus of elasticity and tensile strength[12].

    2.2 Positive interactions betweeen packages and products. Figure 1 shows that polyethylene films absorbed and transferred undesirable oxidative off flavors from Mazzot to the external atmosphere, whereas high barrier packages retained these off odors in the package.

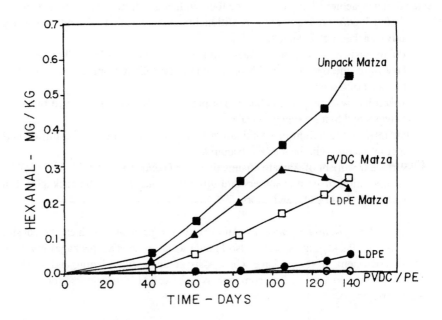

Figure 1 *Effect of package type on hexanal concentration in Mazzot.*

samples at every sensory evaluation time.

2.5 Volatile components of the head space

The package wall was pierced with an injection needle and 50 ml of gas from the head space of the packages passed through the Tenax absorbent. A Varian 3400 gas chromatograph was used to separate the volatile compounds. The phase of the column used was a crosslinked methyl silicone phase, the length was 50 m, diameter 0.2 mm and the film thickness was 0.5 μm. The oven temperature was programmed so that the temperature was at 0 °C 3 min, increased 5 °C/min to 50 °C and 25 °C/min to 240 °C, and held at that temperature for 4.5 min. Volatile compounds were identified with a Finnigan Incos 50 mass spectrometer. (The grammes can not be directly compared quantitatively with one another, because the y-axis of the grammes was set in accordance with the highest peak.)

3 RESULTS AND DISCUSSION

The O_2-concentration increased strongly in the leaking package without an absorbent, but remained low in the packages with an absorbent (Table 1). The CO_2-concentration decreased in all packages at the beginning of the storage when CO_2 was absorbed in the samples (Table 2). However, the decrease was the most remarkable in leaking packages and in packages with an oxygen absorbent. At the end of the storage the CO_2-concentration started to increase, particularly in the packages without an absorbent because of microbial metabolism.

Table 1 *The effect of the leakage, oxygen absorption and storage time on the O_2-concentration in the gas packages of cured cooked ham. The data represents the means of 4 samples ± standard deviation.*

Package type	Storage time (d)/O_2-concentration (%)						
	1	7	10	14	18	23	29
intact + O_2-abs	0.1±0.00	0.5±0.33	0.1±0.00	0.1±0.05	0.2±0.06	0.2±0.05	0.1±0.05
intact	2.3±0.42	2.9±0.20	2.5±0.08	1.1±0.33	0.4±0.17	0.1±0.10	0.2±0.05
leaking* + O_2-abs	1.6±0.32	0.4±0.19	0.3±0.08	0.2±0.19	0.3±0.10	0.3±0.08	0.3±0.06
leaking*	4.4±0.74	7.5±0.65	10.8±0.87	7.3±2.12	6.5±1.88	0.6±0.15	0.4±0.41

Table 2 *The effect of the leakage, oxygen absorption and storage time on the CO_2-concentration in the gas packages of cured cooked ham. The data represents the means of 4 samples ± standard deviation.*

Package type	Storage time (d)/O_2-concentration (%)						
	1	7	10	14	18	23	29
intact + O_2-abs	7±1.1	4±0.5	2±0.5	2±0.5	3±0.8	5±2.6	16±9.0
intact	9±0.5	8±0.0	7±0.0	8±1.0	9±1.5	19±4.3	19±5.5
leaking* + O_2-abs	6±0.6	2±0.6	0±0.0	0±0.0	0±0.5	1±0.5	10±4.0
leaking*	8±0.6	7±0.0	5±0.5	7±1.5	8±2.6	18±3.6	29±2,9

* There was one microhole (diameter 95 μm and length 3mm) in the sealing area.

The package leakage increased the growth rate of moulds and yeasts and *Brochothrix thermosphacta*, while the use of an oxygen absorbent decreased the growth rate of these microbes (Figures 1a and 1b). The lactic acid bacteria, mesophilic and psychrophilic bacteria were unaffected by package leakage and the use of an oxygen absorbent. The pH of the sliced ham increased during the middle of the storage, probably because of a breakdown of proteins. At the end of the storage the pH decreased. This was likely due to the growth of lactic acid bacteria (Figure 1c). The similar changes of the pH of cooked ham packed in modified atmosphere during storage have been reported earlier.[2,3]

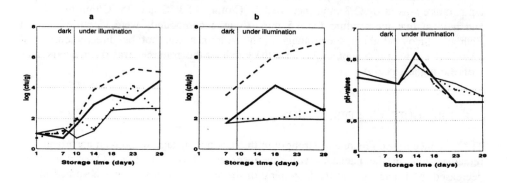

Figure 1 *Effect of the package leakage and oxygen absorption on the number of yeasts and moulds (a),* Brochothrix thermosphacta *(b) and the pH-values (c) of gas-packed ham.* ⸻ *intact package with an O₂-absorbent,* ▬ *intact,* • • *leaking package with an oxygen absorbent,* ⸺ ⸺ *leaking package. The data represent the means of two duplicate samples.*

The sensory quality of ham in the packages without an oxygen absorbent deteriorated rapidly. In these packages, the leakage affected the quality of the ham at the beginning of the storage, but after 14 days there was no difference between the ham in the intact and leaking packages, neither sample was acceptable for human consumption anymore. In the packages with an oxygen absorbent, the ham was still acceptable after 29 days both in the intact and leaking packages (Tables 3 and 4). The colour of ham, measured both by sensory evaluation and with a Minolta Chroma Meter, was highly dependent on the package leakage and the oxygen absorbent. The leakage accelerated the colour change of ham from red to grey, while the oxygen absorbent prevented the colour change. The light together with residual oxygen changes the colour of the ham quickly[4]. In the packages without an oxygen absorbent a sour and spoiled odour developed faster than in the packages with an oxygen absorbent. At the end of the storage the ham in the leaking package without an absorbent was described as having a fermented odour.

Table 3 *Effect of the leakage, oxygen absorption and storage time on the sensory quality of gas-packed cured cooked ham. The scale used was a quality scale from 0 to 5, (0 = unfit, 5 = excellent). O = odour and A = appearance of the product. The same letter after the mean means that there is no significant difference between the samples. The data represents the means of 4 - 6 panelists ± standard deviation.*

Package type	1 d		7 d		10 d	
	O	A	O	A	O	A
refe-rence	4,9[a] ±0,25	5,0[a] ±0,00	4,7[a] ±0,45	4,8[a] ±0,45	4,8[a] ±0,24	4,9[a] ±0,13
intact + O_2-abs	4,6[a] ±0,32	4,9[a] ±0,25	4,2[a] ±0,33	4,7[a] ±0,45	3,8[b] ±0,24	4,3[a] ±0,32
intact	4,5[a] ±0,17	4,9[a] ±0,25	4,1[a] ±0,68	4,7[a] ±0,45	4,0[ab] ±0,41	4,5[a] ±0,46
leaking +O_2-abs	4,6[a] ±0,24	4,9[a] ±0,25	4,1[a] ±0,48	4,7[a] ±0,45	3,8[b] ±0,32	4,3[a] ±0,20
leaking	4,6[a] ±0,13	4,9[a] ±0,25	4,1[a] ±0,63	4,5[a] ±0,87	3,3[b] ±0,69	2,9[b] ±0,59

Table 3 (continued).

Package type	14 d		18 d		23 d		29 d	
	O	A	O	A	O	A	O	A
refe-rence	4,7[a] ±0,42	4,7[a] ±0,41	4,6[a] ±0,38	4,8[a] ±0,45	4,7[a] ±0,45	4,8[a] ±0,21	4,7[a] ±0,47	4,7[a] ±0,47
intact + O_2-abs	3,6[bc] ±0,54	4,2[a] ±0,67	3,4[b] ±0,38	3,9[ab] ±0,89	3,5[b] ±0,42	4,1[a] ±0,67	2,9[b] ±0,33	3,0[ab] ±1,08
intact	1,9[e] ±0,54	1,2[c] ±0,27	1,3[c] ±0,74	0,8[d] ±0,50	2,4[c] ±0,51	0,6[d] ±0,52	1,6[bc] ±0,66	1,3[b] ±0,87
leaking +O_2-abs	2,6[d] ±0,61	2,6[b] ±0,82	3,0[b] ±0,34	3,2[bc] ±0,74	2,6[c] ±0,66	2,0[c] ±0,35	2,9[b] ±0,47	2,5[ab] ±1,78
leaking	1,0[f] ±0,53	1,1[c] ±0,22	0,7[d] ±0,87	0,6[d] ±0,45	1,0[d] ±0,54	0,6[d] ±0,45	1,0[c] ±0,68	0,9[b] ±1,03

The composition of the volatiles in the head-space of packages was also affected by the leakage and the oxygen absorbent. The amount of ethanol was greater in the head-space of leaking packages than in that of the intact packages. In the head-space of the intact and leaking packages with an oxygen absorbent the amount of ethanol was smaller than in the head-space of packages without an absorbent (Figure 2). Many bacteria have been found to produce ethanol[5] and ethanol has been proposed as a potential indicator of spoilage of canned meat[6] and vacuum-packed fish[7]. On the basis of the results of this study, ethanol appears to be a potential quality indicator of gas packed ham.

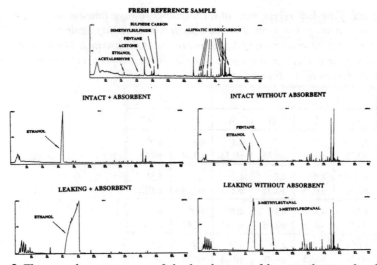

Figure 2 *The gas chromatograms of the head-space of ham packages after 29 days of storage.*

4 CONCLUSION

The light and oxygen together spoiled ham very quickly. The quality of ham stayed good surprisingly long even in leaking packages, when an oxygen absorbent was used. Actually, in leaking packages with an oxygen absorbent, quality stayed better than in intact packages without an oxygen absorbent. Leakage affected the quality of ham significantly only after 23 days of storage, when the oxygen absorbent was used. The oxygen absorbent together with modified atmosphere packaging assured better shelf-life for ham than modified atmosphere packaging without an absorbent.

References

1. R.A. Lampi, *J. Food Process. Eng.*, 1980, **4**, 1.

2. H. Silla and B. Simonsen, *Fleischwirtsch.*, 1985, **65**, 116.

3. R. Ahvenainen, E. Skyttä, and R.-L. Kivikataja., *Lebensm. -Wiss. u. -Technol.*, 1985, **23**, 139.

4. H.J. Andersen, G., Bertelsen, L., Boegh-Soerensen, C.K., Shek and L.H. Skibsted., *Meat Sci.*, 1988, **22**, 283.

5. A. Ahamed and J.R. Matches., *J. Food Prot.*, 1993, **46**, 1055.

6. M.J. Eyles and R.F. Adams., *Int. J. Food Microbiol.*, 1986, **3**, 321.

7. H. Rehbein., *Arch. Lebensm.*, 1993, **44**, 1.

Colour of Beef Loins Stored in Carbon Dioxide with Oxygen Scavengers

O. Sørheim, P. Lea, A. K. Arnesen, and J. Haugdal

MATFORSK - NORWEGIAN FOOD RESEARCH INSTITUTE, OSLOVEIEN 1, N-1430 ÅS, NORWAY

1 INTRODUCTION

Modified atmosphere packaging (MAP) is increasingly used for extending the shelf-life of chilled beef and other types of meat. The shelf-life of meat can be limited by either microbiological spoilage or deterioration of colour. Storage in atmospheres containing carbon dioxide (CO_2) reduces the growth of many microorganisms. Unfortunately, CO_2 is claimed to discolour meat, in particular at concentrations above 20 %[1,2]. However, there are conflicting data on the effect of CO_2, as other researchers found no discoloration of beef stored in CO_2[3,4].

The colour of meat is known to be sensitive to low concentrations of oxygen (O_2), due to formation of gray or brown metmyoglobin[1]. Beef developed noticeable browning with more than 0.15 % O_2 in a CO_2 atmosphere[4]. Oxygen scavenging systems can reduce or eliminate the metmyoglobin formation in meat by removing O_2 from the atmospheres[3,5]. The objective of the present study was to compare the colour stability of beef stored in vacuum or in CO_2 with commercial O_2 scavengers, both at two different chilling temperatures.

2 MATERIALS AND METHODS

2.1 Meat Samples

Twelve beef loins were obtained at a commercial abattoir 3 days after slaughter. The carcasses of Norwegian Red Cattle had been electrically stimulated, were graded lean or extra lean and were without dark, dry and firm meat. The loins were deboned and four sections of approximately 500 g were cut from the <u>longissimus lumborum</u> of each loin. The sections were assigned to the following packaging and storage conditions:
* vacuum at 2 °C (V2)
* vacuum at 6 °C (V6)
* CO_2 at 2 °C (M2)
* CO_2 at 6 °C (M6).

2.2 Packaging and Storage

Within 15 minutes after cutting, the sections were packaged on an Intevac chamber machine (Intevac Verpackungsmaschinen, Wallenhorst, Germany). The bags were of type polyamide/polyethylene (Halvorsen & Larsen A/S, Oslo) with an O_2 transmission rate of 30 $cm^3/m^2/24$ h/atm at 23 °C and 0 % RH. The vacuum cycle was 22 seconds. Initial CO_2 to meat ratio was more than 1.5:1 by volume. One O_2 scavenger of type Ageless FX-100 ® (Mitsubishi Gas Chem. Co. Inc., Tokyo, Japan) was inserted in each bag with CO_2. All samples were stored in the dark for 31 days.

2.3 Gas Analyses

Oxygen was analysed with a Toray LC-700F and CO_2 with a Toray PG-100 (both Toray Eng., Japan) immediately after packaging and at days 1, 2, 3 and 31 of storage.

2.4 Instrumental Colour Analysis

A Minolta Chroma Meter CR-200 (Minolta Camera Co., Osaka, Japan) with 8 mm viewing port and illuminant D_{65} was used for measuring CIE(1976) L* (lightness), a* (redness) and b* (yellowness) values[6] on the sections. The measurements were made through the packaging film at days 1, 3, 7, 14, 21 and 31 of storage. The values were corrected for those of the packaging film by subtraction.

2.5 Sensory Colour Evaluation

A five member trained panel evaluated three sensory colour characteristics of the sections after 31 days storage. The characteristics were: colour (scale 1 = bright red, 2 = dark red, 3 = slightly gray/brown, 4 = moderately gray/brown and 5 = extremely gray/brown), worst spot colour (scale as for colour) and discoloration (scale 1 = none, 2 = 1-10 %, 3 = 11-20 %, 4 = 21-60 %, 5 = 61-100 % of the muscle area).

2.6 Microbiology

Total viable counts were analysed on Plate Count Agar (Difco) after sampling on the deboned whole loins and on the sections at the end of storage[5]. Sampling was performed by cutting 25 cm^2 squares of the meat surface.

2.7 Statistics

Analysis of variance with Tukey's multiple comparisons test was performed by using the SAS statistical programme[7].

3 RESULTS AND DISCUSSION

3.1 Gas Composition

The bags with CO_2 had O_2 concentrations below 0.5 % after packaging. After 3 days storage, the O_2 absorbers had consumed all residual O_2 in these bags and no O_2 was detected thereafter. The CO$_2$ concentrations were reduced from initial 98 % to approximately 90 % at the end of storage, due to CO$_2$ absorption in the meat[4,5].

3.2 Microbiology

Total viable counts on the loins before packaging were below 10^3/cm^2. After 31 days storage the total viable counts on the sections had increased to an average of: 5 x 10^5 for M2, 6 x 10^6 for M6, 4 x 10^6 for V2 and 1 x 10^7/cm^2 for V6, with M2 and V6 being significantly different (p < 0.05). The results show that bacterial growth was reduced both by CO$_2$ and low temperature storage.

3.3 Colour

CIE(1976) L*a*b* values of the sections during storage were affected by packaging conditions, but not by temperature; see Figures 1, 2 and 3. In general, the sections in CO$_2$ were more light, less red and more yellow than the sections in vacuum. The sections in CO$_2$ had increased L*, decreased a* and increased b*, while the sections in vacuum also had increased L*, but stable a* and b* during the storage period. The CO$_2$ samples (M2 and M6) were significantly different (p < 0.05) from the vacuum samples (V2 and V6) in L* after one day, in a* after 7 days and in b* after 14 days storage.

The colour differences were not as evident when evaluated by the sensory panel as when measured instrumentally; see Table 1. The sensory results showed that the M6 sections had higher score on worst spot colour and were more discoloured than the V2 sections (p < 0.05). The data of the sensory evaluation seem to correspond most with the a* of the CIE (1976) L*a*b* values.

Before packaging, fresh cut meat exposed to air usually has a layer of bright red oxymyoglobin on the surface. Initially in a typical CO$_2$ storage, low concentrations of O$_2$ are expected to facilitate gray/brown metmyoglobin formation. However, when all the O$_2$ is consumed e.g. by O$_2$ scavengers, the metmyoglobin should change into dark red deoxymyoglobin, because of reducing activity in the muscle[8]. The process of muscle pigment convertion in CO$_2$ packaged beef may take up to 48 hours[9,10]. The slight increase

Table 1   *Sensory Evaluation of Beef Loin Sections after 31 Days of Storage*

Storage Conditions	Colour[1]	Worst Spot[1] Colour	Discoloration[2]
CO$_2$ at 2 °C	2.2^a	2.3^{a,b}	1.4^{a,b}
CO$_2$ at 6 °C	2.2^a	2.6^a	1.9^a
Vacuum at 2 °C	2.0^a	2.0^b	1.0^b
Vacuum at 6 °C	2.1^a	2.4^{a,b}	1.3^{a,b}

[1]) Scale: 1 = bright red to 5 = extremely gray/brown
[2]) Scale: 1 = none to 5 = 61-100 % of muscle area
[a,b] Means in a column without common superscript letters are significantly different (p < 0.05)

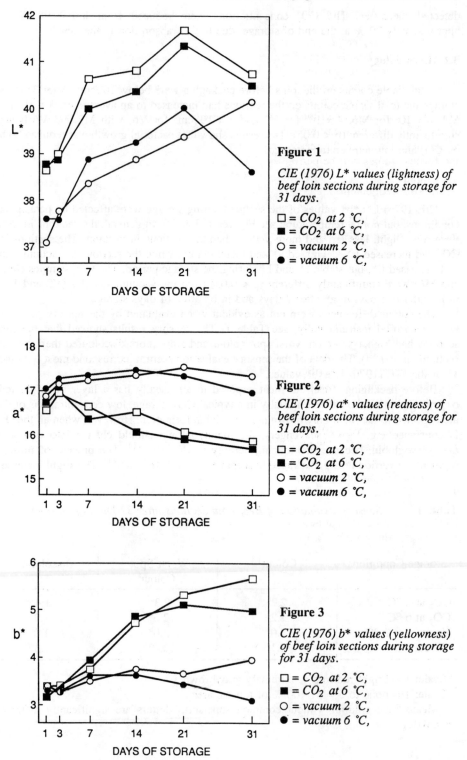

Figure 1

CIE (1976) L* values (lightness) of beef loin sections during storage for 31 days.

□ = CO₂ at 2 ℃,
■ = CO₂ at 6 ℃,
○ = vacuum 2 ℃,
● = vacuum 6 ℃,

Figure 2

CIE (1976) a* values (redness) of beef loin sections during storage for 31 days.

□ = CO₂ at 2 ℃,
■ = CO₂ at 6 ℃,
○ = vacuum 2 ℃,
● = vacuum 6 ℃,

Figure 3

CIE (1976) b* values (yellowness) of beef loin sections during storage for 31 days.

□ = CO₂ at 2 ℃,
■ = CO₂ at 6 ℃,
○ = vacuum 2 ℃,
● = vacuum 6 ℃,

in a* on the sections from day 1 to 3 of storage indicates conversion of pigment; see Figure 2.

The slight graying/browning measured and observed on the beef loin sections treated with CO_2, is not readily explained by the data of this experiment. Bacterial growth was not likely to cause the discoloration, because the total viable counts were lower on the CO_2 samples than the vacuum samples. The exposure to low concentrations of O_2 early in the storage period may possibly damage the colour later during the storage. Finally, the CO_2 per se can have a detrimental effect on the colour of beef[1,2]. A weak pH depression of about 0.1 pH unit measured in beef exposed to CO_2[11] may reduce the colour stability of the meat. However, the reason why CO_2 may affect the colour of beef is still unclear and further studies will be performed to clarify this issue.

Acknowledgements

The authors wish to thank Dr Hilde Nissen and Therese Hagtvedt for the microbiological analyses. We are grateful for the gift of Ageless ® O_2 scavengers from W. R. Grace A/S, Denmark.

References

1.   D. A. Ledward, *J. Food Sci.*, 1970, **35**, 33.
2.   G. C. Holland, Proc. Meat Ind. Res. Conf., Amer. Meat Inst. Washington D.C., 1980, 21.
3.   M. O'Keeffe and D. E. Hood, *Meat Sci.*, 1980, **5**, 27.
4.   N. Penney and R. G. Bell, *Meat Sci.*, 1993, **33**, 245.
5.   O. Sørheim, J. Aasgaard Grini, H. Nissen, H. J. Andersen and P. Lea, *Fleischwirtsch.*, (in press).
6.   R. S. Hunter and R. W. Harold, "The Measurement of Appearance", John Wiley & Sons, New York, 1987.
7.   SAS Institute Inc., "SAS User's Guide", SAS Institute, Cary, N.C., USA, 1987.
8.   D. H. Kropf, M. C. Hunt and D. Piske, Proc. Meat Ind. Res. Conf., Amer. Meat Inst., Washington D.C., 1986, 62.
9.   M. D. Pierson, D. L. Collins-Thompson and Z. J. Ordal, *Food Technol.*, 1971, **24**, 129.
10.   C. O. Gill and T. Jones, *Meat Sci.*, 1994, **37**, 281.
11.   D. L. Huffman, K. A. Davis, D. N. Marple and J. A. McGuire, *J. Food Sci.*, 1975, **40**, 1229.

Reducing Package Deformation and Increasing Filling Degree in Packages of Cod Fillets in CO_2-enriched Atmospheres by Adding Sodium Carbonate and Citric Acid to an Exudate Absorber

Bjørn Bjerkeng, Morten Sivertsvik, Jan Thomas Rosnes, and Helge Bergslien

NORCONSERV, INSTITUTE OF FISH PROCESSING AND PRESERVATION TECHNOLOGY, PO BOX 327, N-4001 STAVANGER, NORWAY

1 INTRODUCTION

The preservative effect of high CO_2 levels in the surrounding atmosphere of fish products during storage, was originally exploited in the 1930's in the UK (e.g. Coyne[1,2]). Modified atmosphere (MA) packaging of fish and fish products have recently been reviewed by several authors.[3-6] Packaging of fish products in atmospheres with elevated levels of CO_2 has been disadvantaged by the high solubility of CO_2 in fish muscle. This CO_2 absorption leads to gas volume contraction, changes the initial gas composition considerably during storage, and can result in package deformation or collapse,[6] as illustrated in Figure 1. Currently, the manufacture of MA packaged fish products is hampered by a high ratio between gas and product volume (G/P), causing high costs of packaging and transportation. Recently, Labuza[7] suggested the incorporation of a CO_2 generating sachet system in packages of foods benefiting from high levels of CO_2. The aim of the present study was to develop a packaging method yielding low G/P ratios, which could provide a high and stable CO_2 level in the gas phase during storage, and reduce the risk for package collapse, by development of CO_2 from a mixture of sodium carbonate and citric acid inside the package.

Figure 1 *Illustration of MA package appearance. a) Contraction due to CO_2 absorption, b) Package with CO_2-generator (mixture of sodium carbonate and citric acid).*

2 EXPERIMENTAL

2.1 Materials

Cod (*Gadus morhua*, weight 0.8-1.5 kg) was slaughtered, filleted and packaged within four hours *post mortem*. Portioned fish (300±1g) was packaged in high density polyethylene (HDPE) trays (410 ml, Dynopack tray 80523, Dynopack, Kristiansand, Norway) and subjected to three different treatments. Exudate absorbers containing a mixture of food grade sodium carbonate and citric acid (1:1) were immediately placed under the cod fillets. The exudate absorbers consisted of eight layers of paper wad (8cm x 11cm) capable of absorbing about 7g of water or exudate. In treatment 1 the absorbers contained 4.0g of the carbonate mixture, and the packages were MA packaged (initial gas composition 68.8% CO_2, 29.4% N_2 and 1.8% O_2). For treatment 2, the cod fillets were dipped in a 20% solution of sodium chloride for 20s before packaging as for treatment 1. For treatment 3 the absorbers contained 6.0g of the carbonate mixture and the packages were vacuumated. Cod fillets packaged with exudate absorbers and subjected to MA packaging as for treatment 1 were used as a control.

2.2 Packaging and Storage

The packaging of treatments 1, 2 and the control group was accomplished by evacuating the HDPE trays and backflushing with a mixture of 70% CO_2 and 30% N_2 (HydroGas, Oslo, Norway) using a packaging machine (Dyno 462 VGA, Dynopack, Kristiansand, Norway) with heat-sealing of a 90μm PE/PA-foil as lidding material. The vacuum packaging in treatment 3 was accomplished similarly without backflushing.

The cod fillets were stored in darkness at 3°C and samples were collected for analyses after 0, 1, 4, 7, 11 and 15 days of storage, respectively.

2.3 Chemical Analyses

The results presented for head-space gas composition, fish fillet surface pH and formation of exudate are averaged for three samples, while trimethylamine content are averaged from two measurements. The results are presented with standard deviations.

2.3.1 Head-space gas composition. The head-space gas composition in the packages was determined in triplicate by injecting an aliquot (30ml) of the head-space gas of the trays using an oxygen and carbondioxide analyzer (MAP Test 2000, Hitech Instruments, Luton, UK). The gas was collected with a syringe after intrusion of the top foil. The analyzer was calibrated against a commercial gas mixture (O_2:CO_2:N_2 (1.09:44.1:54.8); Norsk Hydro, Rjukan, Norway) and air before each sampling. The diffusion of gases through the packaging materials was investigated by monitoring changes in gas composition of empty trays containing the gas mixture employed in treatment 1, during storage at 3°C.

2.3.2 Fish fillet surface pH. The pH of the fish fillet surface was determined in triplicate using a surface electrode (combined pH electrode GK2501, Radiometer, Copenhagen, Denmark) connected to an Unicam 9450 pH meter (Unicam, Cambridge, UK).

2.3.3 Formation of exudate. Formation of exudate in the packages was determined in triplicate as difference in weight between trays before and after removal of the fish fillets. Exudate formation is expressed as % of initial fillet weight.

2.3.4 Trimethylamine content. Trimethylamine (TMA) was determined in duplicate using a modified Conway micro-determination method.[8] The homogenized fish fillets (25g) were extracted with trichloroacetic acid (7.5%, 75ml) on a shaking machine for 30min. An aliquot of the filtered extract (2ml) was added to the outer chamber of a Conway and allowed to react with a methanal solution (40%, 1ml) neutralized with $MgCO_3$, in order to transform any ammonia, mono- and diamines present. The mixture was reacted for 5min under a glass cover. The inner chamber of the Conway dish contained a boric acid solution (2ml, 10g H_3BO_3 in 210ml 96% ethanol and 700ml dist. water) neutralized with NaOH. Bromocresol green and methyl red was added to the boric acid solution as indicators. A saturated solution of K_2CO_3 (2ml) was added to the outer chamber and mixed by gently swaying the Conway dish. TMA was allowed to diffuse overnight at ambient temperature. The TMA content was calculated from the amount of 0.0143N HCl required to neutralize the solution of the inner chamber of the Conway dish. The results are expressed as mg Nitrogen/100g fish.

2.4 Microbiological Analyses

Total viable counts (TVC) and H_2S-producing bacteria were determined in triplicate as pour plate counts after incubation on Agar Lyngby (IA) (Oxoid CM 867), 20°C for 3 days. Psychrophilic bacteria were determined as spread plate counts (plate count agar), 8°C for 4-6 days. All plates were incubated in air. Determination of shelf-life was based on sensory evaluation of ammonia-like smell from the fillets. Averaged results are presented as log (colony forming units (CFU))/g fish fillet.

3 RESULTS AND DISCUSSION

When the (1:1)-mixture of sodium carbonate and citric acid employed in this experiment, was soaked with fish fillet exudate, CO_2 gas was developed. Although only sodium carbonate and citric acid was tested, a number of other carbonates and acids may be used. Optimization of the ratio between the carbonate and acid, and the rate of formation of CO_2 was not attempted in this experiment. Such an optimization is likely to depend on the amount of exudate (fish freshness), pH, storage temperature, the volume of head-space and the amount of CO_2 absorbed by the fish.

A content of CO_2 gas higher than 50% was maintained in the initially vacuumated packages (treatment 3) from day one and throughout the experiment, Figure 2. The CO_2 content of the gas packaged samples containing the carbonate mixture (treatment 1 and 2) decreased to approximately 40% after 4 and 7 days of storage, respectively, while the

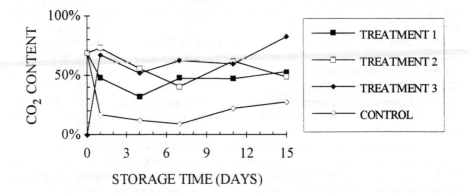

Figure 2 *Development of head-space CO_2 content during chilled storage of cod fillets packaged in different CO_2-enriched atmospheres.*

CO_2 level in the control group rapidly decreased to below 20% after storage for one day. Packages with the carbonate mixture maintained their initial shape, while dissolution of CO_2 gas into the fish fillets in the control group led to package deformation, as illustrated in Figure 1. Diffusion of gases through the packaging materials was negligible.

The growth of selected bacterial groups (psychrophilic and H_2S-producing bacteria and TVC) are shown in Figure 3. No relevant increase in TVC was observed for treatment 1 and 2, while there were some increase in treatment 3 and control. The TVC was not reflected by the formation of TMA, Table 1, or sensory evaluation. Microbial shelf life was therefore based on H_2S-producing and psychrophilic bacteria. The H_2S-producing *Shewanella putrefaciens* has been identified as the main spoilage bacteria of whole cod stored in ice and of chilled vacuum packaged cod fillets.[9,10] Growth rate and maximum concentration of H_2S-producing bacteria are reduced when cod is stored in MA with increased CO_2-levels.[11,12] For the control group, with the lowest CO_2-level, numbers of psychrophilic and H_2S-producing bacteria increased with about 2 log_{10}-units during the first 7 days of storage. In contrast, all treatments had higher CO_2-levels, in addition to either lower growth rates or lower bacterial numbers. The highest bacterial numbers and growth rates were observed for psychrophilic bacteria. This is consistent with earlier reports,[12,13] suggesting that the main spoilage organisms are marine vibrios/*Photobacterium phosphoreum* growing at high CO_2 levels at chilled temperatures. When the microbial shelf-life was determined as time needed for psychrophilic bacteria to reach log CFU/g=6, the shelf-life of the initially vacuumated cod fillets was increased approximately 3 days

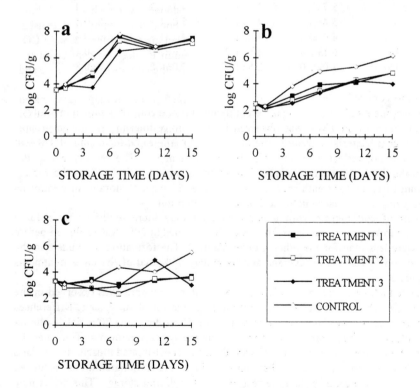

Figure 3 *Development of a) psychrophilic and b) H_2S-producing bacteria, and c) TVC during storage (3°C) in CO_2-enriched atmospheres.*

Table 1 *Development of TMA During Chilled Storage of Cod Fillets Packaged in Different CO_2-Enriched Atmospheres.*

Storage time/	$TMA/10^{-2}$ mg-N g^{-1}			
days	Treatment 1	Treatment 2	Treatment 3	Control
0	0	0	0	0
1	0	0	0	0
4	0.4±0	0.8±0	0.6±0	3.6±0.3
7	8.3±2.1	29.5±0.1	4.6±0.3	32.4±1.1
11	nd[a]	nd	nd	nd
15	81.2±0	85.0±0.3	74.3±1.6	77.8±0.6

[a] nd = not determined

Table 2 *Development of Exudate During Chilled Storage of Cod Fillets Packaged in Different CO_2-Enriched Atmospheres.*

Storage time/	Exudate/% of initial fish weight			
days	Treatment 1	Treatment 2	Treatment 3	Control
0	0	0	0	0
1	5.7±1.5	3.6±1.2	4.4±1.3	4.5±1.4
4	5.4±0.5	4.7±1.7	5.0±1.1	3.8±0.6
7	4.9±0.8	4.0±1.5	4.2±1.1	4.8±0.4
11	6.1±1.4	3.7±1.2	5.6±2.0	5.4±0.1
15	3.0±1.0	2.8±1.7	5.0±0.8	6.0±0.8

as compared to control.

The development of TMA for the different treatments and control is shown in Table 1. A more rapid increase in TMA was observed in the control than in the treated groups, indicating a higher bacterial activity in the former group. The rapid formation of TMA in treatment 2 was not reflected by the microbiological analyses. The sensory evaluation showed that the fillets were acceptable for human consumption until day 7 for the control group and until day 11 for treatments 1, 2 and 3. After 15 days of storage all treatments and control group were unacceptable to human consumption.

Treatment of cod muscle with a NaCl solution may increase the water holding capacity.[14] The formation of exudates from cod fillets dipped in 20% NaCl solution before packaging were less than for the other groups, Table 2. The formation of exudate in the NaCl dipped cod fillets was on average approximately 75% that of the other treatments and the control.

A small reduction of the initial pH (ca. 6.8) of about 0.2 units was observed in the treated packages. This difference evened out after storage for about 7 days. Differences in fillet colour were observed between treatments. The fillets dipped in 20% NaCl-solution obtained a bluish surface colour, not observed in the other treatments. This may be attributed to the increased transparency observed for the NaCl-treated muscle tissue. Parts of the fillet in direct contact with an absorber containing the carbonate mixture achieved a noticeably whiter appearance than the rest of the fillet during storage. This effect may be caused by the denaturation of surface proteins caused by the contact with the citric acid. A sachet system avoiding this contact may provide a more homogenous fish fillet appearance.

4 CONCLUSION

We have presented preliminary results concerning reduction of package deformation during MA packaging of fish. This reduction was achieved by adding an exudate absorber containing a mixture of sodium carbonate and citric acid inside the package. High and stable CO_2 levels (>50%) were obtained during storage at 3°C, increasing the shelf-life of cod fillets for about three days.

Based on these results, decreased G/P ratios compared to the currently employed ratios may be used, without occurrence of possible package deformation during MA packaging of fish. The future development of a sachet system generating CO_2 may also eliminate quality deteriorations due to contact between product and any acid present. The use of a CO_2-generating system in initially vacuumated packages may entirely eliminate the use of gases in products where high CO_2 levels are preferred.

Dipping the fish fillets in 20% NaCl solution before MA packaging reduced exudate formation, but an unattractive bluish appearance resulted.

Acknowledgement

This work was kindly supported by the Norwegian research programme, NÆRFORSK (grant no 267816). The technical staff at Norconserv are gratefully acknowledged for their skillful assistance.

References

1. F. P. Coyne, *J. Soc. Chem. Ind.*, 1932, **51**, 119T.
2. F. P. Coyne, *J. Soc. Chem. Ind.*, 1933, **52**, 19T.
3. A. Pedrosa-Menabrito and J. M. Regenstein, *J. Food Qual.*, 1990, **13**, 129.
4. C. Pellegrino, V. Giaccone and E. Parisi, *Fleischwirtschaft*, 1990, **70**, 1340.
5. B. J. Skura, In 'Modified Atmosphere Packaging of Food', B. Ooraikul and M. E. Stiles (ed.), Ellis Horwood, New York NY, USA, 1991, Chapter 6, p. 148.
6. N. R. Reddy, D. J. Armstrong, E. J. Rhodehamel and D. A. Kautter, *J. Food Safety*, 1992, **12**, 87.
7. T. P. Labuza, In 'Science for the Food Industry of the 21st Century', M. Yalpani (ed.), ATL Press, Mount Prospect IL, USA, 1993, Chapter 17, p. 265.
8. E. J. Conway and A. Byrne, *Biochem. J.*, 1933, **27**, 419.
9. L. Gram, G. Trolle and H. H. Huss, *Int. J. Food Microbiol.*, 1987, **4**, 65.
10. B. R. Jørgensen and H. H. Huss, 1989, *Int. J. Food Microbiol.*, 1989, **6**, 295.
11. H. Einarsson, In 'Quality Assurance in the Fish Industry', H. H. Huss, M. Jakobsen and J. Liston (ed.), Elsevier, London, UK, 1992, p. 479.
12. P. Dalgaard, L. Gram and H. H. Huss, *Int. J. Food Microbiol.*, 1993, **19**, 283.
13. P. Dalgaard, PhD Thesis, Technical University, Lyngby, Denmark, 1993.
14. R. Schubring and E. Sandau, *Fisch. Forsch. Rostock*, 1989, **27**, 5.

Subject Index

New and Recent books on
FOOD SCIENCE AND NUTRITION

Food Macromolecules and Colloids
Edited by E. Dickinson, University of Leeds, UK
D. Lorient, Université de Bourgogne, France

Hardcover   Approx 380 pages
ISBN 0 85404 700 X   1995   Price £92.50

Food: The Definitive Guide
By T. Coultate and J. Davies, South Bank University

Softcover   v + 167 pages
ISBN 0 85186 431 7   1994   Price £12.50

Vitamin C - Its Chemistry and Biochemistry
By Michael B. Davies, John Austin and David A. Partridge Anglia
Polytechnic University, Cambridge, UK

Softcover   x + 154 pages
ISBN 0 85186 333 7   1991   Price £13.50

Food: The Chemistry of Its Components
2nd Edition 4th Reprint 1995
By T.P. Coultate, South Bank University

Softcover   xii + 326 pages
ISBN 0 85186 433 3   1989   Price £11.95

Aluminium in Food and the Environment
3rd Reprint 1994
Edited by Robert C. Massey, Ministry of Agriculture, Fisheries and Food,
Norwich
David Taylor, Imperial Chemical Industries PLC, Brixham

Softcover   viii + 108 pages
ISBN 0 85186 846 0   1989   Price £27.50

Prices subject to change without notice.

To order please contact:
Turpin Distribution Services Ltd., Blackhorse Road, Letchworth, Herts SG6 1HN, UK.
Tel: +44 (0) 1462 672555. Fax: +44 (0) 1462 480947.

For further information please contact:
Sales and Promotion Department, Royal Society of Chemistry,
Thomas Graham House, Science Park, Milton Road, Cambridge CB4 4WF, UK.
Tel: +44 (0) 1223 420066. Fax: +44 (0) 1223 423623. E-mail: (Internet) RSC@RSC.ORG.

RSC members are entitled to a discount on most RSC products, and should contact
Membership Administration at our Cambridge address.

THE ROYAL
SOCIETY OF
CHEMISTRY

Information
Services

New and Recent books on
FOOD SCIENCE AND NUTRITION

Maillard Reactions in Chemistry, Food, and Health
Edited by T.P. Labuza and G. A. Reineccius, University of Minnesota, USA
V. Monnier, Case Western Reserve University, USA
J. O'Brien, University College Cork, Republic of Ireland
J. Baynes, University of South Carolina, USA

Hardcover   xviii + 440 pages   ISBN 0 85186 802 9   1994   Price £67.50

Food Microbiology
By M. R. Adams and M. O. Moss, University of Surrey

Softcover   390 pages   ISBN 0 85404 509 0   1995   Price £22.50

Food and Cancer Prevention: Chemical and Biological Aspects
Edited by K. Waldron, AFRC; I.T. Johnson and G.R. Fenwick,
Institute of Food Research, Norwich

Hardcover   xvi + 462   ISBN 0 85186 455 4   1993   Price £59.50

Biochemistry of Milk Products
Edited by A. T. Andrews, Cardiff Institute of Higher Education
J. Varley, University of Reading

Hardcover   viii + 182 pages   ISBN 0 85186 702 2   1994   Price £39.50

Sugarless - Towards the Year 2000
Edited by A.J. Rugg-Gunn, University of Newcastle upon Tyne

Hardcover   x + 198 pages   ISBN 0 85186 495 3   1994   Price £37.00

Developments in the Analysis of Lipids
Edited by J. H. P. Tyman, Brunel University; M. H. Gordon, University of Reading

Hardcover   x + 206 pages   ISBN 0 85186 971 8   1994   Price £45.00

Plant Polymeric Carbohydrates
Edited by F. Meuser, Institute of Food and Fermentation Technology, Berlin, Germany
D.J. Manners, Heriot-Watt University, Edinburgh, UK
W. Seibel, Federal Research Centre for Cereal Potato, and Lipid Research,
Detmold and Münster, Germany

Hardcover   xii + 296 pages   ISBN 0 85186 645 X   1993   Price £57.50

Prices subject to change without notice.

To order please contact:
Turpin Distribution Services Ltd., Blackhorse Road, Letchworth, Herts SG6 1HN, UK.
Tel: +44 (0) 1462 672555. Fax: +44 (0) 1462 480947.

For further information please contact:
Sales and Promotion Department, Royal Society of Chemistry,
Thomas Graham House, Science Park, Milton Road, Cambridge CB4 4WF, UK.
Tel: +44 (0) 1223 420066. Fax: +44 (0) 1223 423623. E-mail: (Internet) RSC@RSC.ORG.

RSC members are entitled to a discount on most RSC products, and should contact
Membership Administration at our Cambridge address.

THE ROYAL
SOCIETY OF
CHEMISTRY

Information
Services